Consumer Behaviors
That Influence
Purchases of Replicate
Entertainment Products

Also by authors

Strategies Required by Managers to Inhibit Movie Piracy (N. Akaeze)

Small Business Sustainability Strategies in Competitive Environments: Small Business and Competition (C. Akaeze)

Consumer Behaviors That Influence Purchases of Replicate Entertainment Products

Dr. Chris Akaeze
and
Dr. Nana Akaeze

Library of Congress Control Number: 2016911300
ISBN: Hardcover   978-1-5245-2465-4
                Softcover   978-1-5245-2464-7
                eBook   978-1-5245-2463-0

Print information available on the last page.

Rev. date: 07/13/2016

To order additional copies of this book, contact:
Xlibris
1-888-795-4274
www.Xlibris.com
Orders@Xlibris.com
746291

CONTENTS

LIST OF TABLES

LIST OF FIGURES

ABSTRACT

Product replication is a growing problem for the entertainment industry and negatively affects multiple affiliate industry. Replication of products costs U.S. movie studios approximately $6 billion annually. The costs of product replication towards creative talents and innovation are significant affecting the economic benefits for original creators of entertainment products and ideas. Guided by Theory of Planned Behaviors, the purpose of this multiple case study was to explore some consumer behaviors that influence complaisance towards purchasing replicate entertainment products in New York City. Data were collected through closed ended qualitative questionnaires from 50 participants who have purchased replicate entertainment products for up to 2 years. Data analysis entailed using thematic analysis. Audit trailing and peer review was used to strengthen the credibility and trustworthiness of the interpretation of participants' responses. The 3 themes that emerged in final report related to Personal Influence, Cultural Influence, and Social Influence towards entertainment consumers purchases of replicate products. Findings from this study may contribute to social change by indicating some consumer behaviors that influence purchases of replicate products. The findings may facilitate strategies for managers to curb replication and mitigate harmful effects to sales and revenue of entertainment products. Data from this study may contribute to the prosperity of entertainment managers, their employees, and local communities. The beneficiaries of this research include entertainment managers, practitioners, academics and policy makers.

DEDICATION

We dedicate this study to our wonderful children Christian II, Solomon, and Samuel, whose encouragement has been significant part of our life pursuits and in particular this study. We also dedicate this study to Mrs. Rabiatu Stella who took the bet on Dr. Nana and commited a lifelong goal to see her educated as much as possible. This is the base of our solid union almost worthy of bylines.

SECTION 1

Foundation of the Study

The Home Entertainment Industry has two major divisions, Hollywood studios with international music corporations and independent players (Oestreicher, 2011). Connaughton and Madsen (2011) classified the entertainment industry as motion picture and video production business. Entertainment businesses comprise establishments that engage in production and distribution of motion pictures, videos, television programs, or television commercials (Connaughton & Madsen, 2011). Entertainment leaders include producers, agents, distributors, and talents that produce records, and package the resultant products (Fradley, 2010; Oliver, 2010).

According to Tolbert, Moore, and Wood (2010), the ultimate goal of business enterprise is to earn a profit. Managers of entertainment coporations can only earn profit through the sales of original products (Oestreicher, 2011). However, product replication practices is the imitating, producing, reproducing, distributing, selling, acquiring, or consuming of unauthentic copies of products without the owners authorization (Ho & Weinberg, 2011; Marcum, Higgins, Wolfe, & Ricketts, 2011; Smallridge & Roberts, 2013) negatively affects sales of original entertainment products. Replication of products costs U.S. movie studios approximately $6 billion annually (Ho & Weinberg 2011).

Product replication is a growing problem for the entertainment industry and negatively affects multiple industries (Smallridge & Roberts, 2013; Walls & Harvey, 2010). Entertainment replication was a reason for decline in sales of about 31% music products between 2004 and 2011 (Smallridge & Roberts, 2013). The U.S. government addresses product replication through laws such as the Prioritizing Resources and Organization for Intellectual Property Act (Pro-IP Act). The use of Copyright laws by entertainment leaders allow for severe penalties including felony charges against copyright violation (Spink

1

& Fejes, 2012). However, in spite of the copyright laws, entertainment-product replication persists, with consequent harm caused to the nation's economy and investors.

Some leaders believe that reasons for continued growth of product replication includes the existing demand because of purchases by consumers to satisfy entertainment wants and needs. According to Chinaka (2016), the central idea behind purchasing a product or service is to satisfy needs and wants. The intent of this qualitative multiple case study is to explore some consumer behaviors that influence purchases of replicate entertainment products. The study focus is on consumers of entertainment movies and music products in New York City (NYC).

Background of the Problem

Entertainment industry is vital to survival of the U.S. economy (Kureshi & Sood, 2011). For example, in 2011, the film industry in the Charlotte area of North Carolina sustained 2,453 full-time jobs a standard considerably higher than the average wage level (Connaughton & Madsen, 2011). Entertainment as an economic sector consists of different products and services including motion pictures, television, music, broadcasting, print media, toys, gaming, gambling, sports, and fine arts. According to Fradley (2010), filmed entertainment production is a major product of the entertainment industry. In addition, the main design in field of entertainment is optical disc product, in various formats including compact discs (CDs) and digital video discs (Oestreicher, 2011). However, with the advent of the Internet and development in digital technology, there has been an increase in the practice of entertainment product replication (Al-Rafee & Dashti, 2012; Rybina, 2011). In addition, advancement in digital technology made the detection and prevention of entertainment-product replication difficult.

Entertainment product replication constitutes a significant and continuing problem, approaching epidemic proportion (Ho & Weinberg, 2011; Hollander, 2011). According to Hsiao-Chien and Wang (2012), replicated goods occupy 5-7% of global trade. There are more counterfeiting of films in NYC than anywhere else in the United States (N. Akaeze, 2016). Replication practices affect the U.S. economy and may exacerbate the country's unemployment rate (Kigerl, 2013; Al-Rafee & Dashti, 2012). Smallridge and Roberts (2013) suggested that practices of replicating entertainment products negatively affect multiple other industries. The practices of replication also have harmful effect on local and national economies (Ahmad, 2010).

Further decline in sales may have a negative effect on the economic viability of many entertainment companies. Morris and Higgins (2010) noted that resultant secondary harms from product replication include loss of artistic talent, job loss, and loss of creative motivation by producers. Sales of music products declined approximately 31% between 2004 and 2011 due to product replication (Smallridge & Roberts, 2013). Since 2000, replication of entertainment products was responsible for about 35% decline in music CD unit sales and over 31% decline in revenue for the same period (Spotts, 2010). There is a reduction in profit margin to entertainment coporations and stakeholders due to an increase in products replication (Burke, 2010; Walls & Harvey, 2010). Revenue loses is as a result of purchases of replicate entertainment products which affects the ability of managers to advance business and creativity.

Furthermore, individuals who engage in product replication infringe copyrights, when they exercise exclusive rights without the owner's authorization (N. Akaeze, 2016; Kigerl, 2013; Lyons, 2010). Copyright law is not effective and has failed as a means to curb the practice of product replication (Koster, 2012). In spite of the enforcement of copyright laws, sales of original entertainment compact discs (CD) in the U.S. showed a decline of more than 74% between 2001 and 2010 (Koster, 2012). The production, seling and buying of replicated entertainment products is increasing significantly and at an alarming rate (Hsiao-Chien & Wang, 2012).

To curb the practice of product replication, entertainment industry leaders and stakeholders may need to adopt different approaches to address the problem. According to Mirza and Islam (2013), entertainment consumers access mass varieties of movies, videos, and television shows-on multiple platforms-faster than ever before at their fingertips. Replication of entertainment products is increasing because there are consumers who purchase replicates to satisfy entertainment wants and needs (Wan, Luk, Yau & Tse, 2009). Exploring consumer behaviors that influence purchases of replicate entertainment products may help managers on strategies to inhibit entertainment products replication. Entertainment managers may develop strategies to reduce the global appetite for replicate products with a better understanding of consumer purchasing behavior (Wiedmann, Hennigs & Klarmann, 2012).

Problem Statement

Approximately 37 % of all entertainment CDs that consumers purchased in 2005 was replicate products while the global traffic of replicate equaled up to $4.5 billion (López-cuñat & Martínez-sánchez, 2015). The U.S entertainment industry is losing an estimated three billion dollars in total annual income to purchases of replicate entertainment-product (Walls & Harvey, 2010).

In the U.S., annual losses to replication of entertainment products result to job losses exceeding 71,000 annually (Al-Rafee & Dashti, 2012). The general business problem is demand by some conusumers for replicate entertainment products. The specific business problem is that some consumers purchasing behaviors are complaisant towards replicate entertainment products.

Purpose Statement

The purpose of this qualitative multiple case study is to explore some consumer behaviors that influence complaisance towards purchases of replicate entertainment products. Responses from a convenient sample of up to 50 participants in New York City to an on-line questionnaire aided collection of qualitative data. The participants who are consumers of entertainment music and video products provided answers to questions displayed for up to two weeks on the Monkey Survey web site and self administered. Results of this study may assist entertainment managers with insights to better understand consumer motivation for purchasing replicate entertainment products. In addition, the findings are potential basis for developing strategies with an aim to reduce global appetite for replicate entertainment products and increase revenue. The social impact is a potential that result of study might lead to sustainable entertainment businesses, which in-turn could save jobs in the entertainment industry and affiliated businesses.

Nature of the Study

A researcher can use the qualitative method to understand social situation of a person, group, or organization (Trotter, 2012). Qualitative method is appropriate for this study because qualitative method is flexible and allows open-ended questions, observation, interviews, and analyzing of documents (Hunt, 2014). In addition, researchers can use qualitative method to gain insight into issues, claims, and concerns by identifying views, opinions, and perceptions of participants (Hunt, 2014).

In contrast, researchers use quantitative methods to project research findings onto larger population through an objective process (Castellan, 2010). Romeo (2010) suggested that most quantitative research have common underlying goal of testing relationships in theories. Quantitative method is not appropriate for this study because there are no deductions achieved through statistical analysis. Mixed methods study is a term for mixing qualitative and quantitative data in a single study (Cameron & Molina, 2011; Harrison & Reilly, 2011). Mixed method is not appropriate for this study because data collection is qualitative without quantitative element.

The choices of designs in qualitative method include case study, phenomenology, grounded theory, ethnography, and narrative research alternatives (Yin, 2011; Hunt, 2014). For this study the choice of multiple case designs was to augment external validity and guard against observer bias (Vissak, 2010). Case study design was appropriate for this study because of the use for clarifying findings (Welford, Murphy, & Casey, 2012). In comparison, a researcher could use phenomenological research design to describe structures of experiences gained in order to arrive at a deeper understanding of a given phenomenon (Cigdemoglu, Arslan, & Akay, 2011).

Phenomenological design was not appropriate for current study because the focus was to explore consumer purchasing behaviors and not primarily to relate depth and understanding of subjects' lived experiences (Hays & Wood, 2011). The purpose of a grounded theory design is to generate new theories, beyond descriptions of individual lived experiences (Hunt, 2014; Marshall & Rossman, 2011). The grounded theory was not appropriate for this study because there was no testing of a theory. Ethnographic study entails a lengthy immersion in the study scenario, allowing the researcher to become integrated (Hunt, 2014). The ethnographic design is not appropriate because this design did not involve an immersion of researcher into a topic scenario, permitting researchers to become an integral element in research scenario. Narrative inquiry was not appropriate because for this study there was no accounting or a giving meaning to experiences in story form (McMullen & Braithwaite, 2013).

According to Yin (2011), a qualitative case study design is vital to finding answers to a research question. The goal of this qualitative case study was to determine the behaviors that influence consumer purchases of replicated entertainment products in NYC. Information obtained from this study may benefit entertainment industry leaders, managers and policymakers who protect copyright of owners. Researchers may use qualitative case study designs rather than quantitative designs to explore how participants interpret phenomenon (Saxena, Gupta, & Ruohonen, 2012).

Research Question

The central research question for this study is as follows: What consumer behaviors influence complaisance towards purchases of replicate entertainment products?

Interview Questions

Since the topic deals with unconventional behaviors of entertainment consumers which minimal individuals want association, online questionnaires were used. Up to 50 consumers that purchased entertainment music and movie products for atleast two years were targeted for answers to the interview questionaires (See Appendix A).

Conceptual Framework

Conceptual framework for this study is theory of planned behavior (TPB). Researchers widely use TPB authored by Ajzen in 1985 to describe the practice of purchasing replicate products (Wiedmann et al., 2012). The explosive growth of information technology facilitates increasing frequency of unfair practices (Hinduja, 2012). Purchasing of replicate entertainment product is an unconventional practice that seems harmless, and involves human conduct (Hinduja, 2011). Wiedmann et al., (2012) associated consumer purchase of replicate products with multifaceted reasons which explains replicate consumption using the TPB.

One reason for the purchases of replicate productts by consumers is that many people do not consider the practice as wrong. The TPB theory includes actions and planned behavior that affect intentions and influence attitude of an individual regarding the practice, consequences of behavior, and subjective norms. Researchers use the TPB theory to analyze attitudes, subjective norms, and perceived behavioral control, to predict intention with relatively high accuracy (Wiedmann et al., 2012).

Furthermore, TPB assumes that a person's intention, when combined with perceived behavioral control will help predict behavior with greater accuracy. Scholars use the TPB to determine purchase intentions, particularly people's situational mood (Hinduja, 2012). In the case of replicated products, researchers use mood processes to explain why people are prone to buy replicate products even when the individuals know the unconventionality of their behavior or the lack of post purchase satisfaction with a product of inferior quality (Hinduja, 2012).

Consumers of replicated products try to legitimize their practice and experience through justifications. Consumers develop coping strategies to reduce their dissonance (Hinduja, 2012). These strategies can alter purchase decisions or modify attitudes (Hinduja, 2012). Most studies on replication have been quantitative with minimal researchers using qualitative methods (Ho & Weinberg, 2011). A number of the studies involved use of TPB in researching product replication. A product buyer begins with recognizing a need, then search for information, evaluates alternatives, and makes a purchase (Spear, 2012). To accomplish the goal of curbing replication, researchers must determine the reason why individuals purchase replicate products over original versions (Spear, 2012). Wiedmann et al., (2012), revealed multifaceted reasons why consumers purchase replicated products using the TPB theory.

The concept of the TPB includes understanding of beliefs about resources and opportunities, often viewed as underlying factors influencing planned behavioral control (Muthiani & Wanjau, 2012). Having positive attitudes toward purchasing a product and having social reasons to do so predict intentions of knowingly purchasing counterfeit products, based on the perceived ease or difficulty associated with this behavior. One factor researchers identified as responsible for the product replication practice is pricing (Muthiani & Wanjau, 2012) using the TPB theory. A business can pursue superior performance by either establishing a cost-leadership position or differentiating the business, products and services from competitions (Parnell, 2011; Porter, 1980). The most important behaviors that influence consumer purchases of replicated products are useful to stakeholders of the entertainment corporation.

The explosive growth of information technology accompanies the increasing frequency of unconventional practices (Hinduja, 2011). Purchasing of replicate products is an unconventional practice involving human conduct and seems harmless. Researchers associated consumer purchase of replicate products with multifaceted reasons explaining counterfeit consumption which relates to economic, functional, individual, and social evaluations with the theory of planned behavior (Wiedmann, Hennigs & Klarmann, 2012).

One reason for the prevalence of replicate movie purchasing is that many people do not consider the practice as wrong. The TPB is a widely used theory to explain practice of purchasing replicate products (Wiedmann et al., 2012). Authored by Ajzen in 1985, the TPB is conceptual framework for this study. The TPB indicates that actions and planned behavior affect behavioral intentions and influence attitude of an individual regarding the practice, consequences of behavior, and subjective norms (Wiedmann et al., 2012). Researchers use TPB to analyze attitudes, subjective norms, and perceived behavioral control, and predict intention with relatively high accuracy. The theory assumes that a person's intention, when combined with perceived behavioral control will help predict behavior with greater accuracy.

Scholars also use the TPB to determine purchase intentions, particularly individual's situational mood. In the case of replicate products, mood processes can explain why individuals are prone to purchase products even when they know the unconventionality of their behavior and lack of post purchase satisfaction with a product of low quality. Consumers of pirated products try to legitimize their practice and experience through justifications. They develop coping strategies to reduce their dissonance. These strategies can alter purchase decisions or modify attitudes. The TPB provides a sound theoretical basis to explore behaviors which influence entertainment consumers' purchases of replicate products. Most studies on product replication have been quantitative with few researchers using qualitative methods. Although most studies have been quantitative, a number of the studies used the TPB in researching replication (Hinduja, 2012).

A product buyer begins with the recognition of a need, then search for information, evaluates alternatives, and makes purchase (Spear, 2012). Wiedmann et al. (2012) revealed the multifaceted reasons why consumers purchase pirated products using the TPB.

The concept of TPB facilitates understanding of beliefs about resources and opportunities, often viewed as underlying factors influencing planned behavioral control (Muthiani & Wanjau, 2012). Having positive attitudes toward purchasing a product and having social reasons to do so predict intentions of knowingly purchasing replicate products, based on perceived ease or difficulty associated with behavior. In line with the concept of TPB, Muthiani and Wanjau (2012) identified that a significant factor influencing product replication is pricing.

Definition of Terms

Entertainment product. Entertainment product is an optical disc, which the entertainment producers make in various formats of compact discs and digital video discs (Oestreicher, 2011).

Movie replication. Movie replication is the act of producing, acquiring, or consuming unlicensed copies of any authentic movie products (Ho & Weinberg, 2011).

Product Replication. In general exists in four common forms of intellectual property rights (IPRs) infringements which includes; counterfeiting, piracy, imitation and gray market (Wiedmann, Hennigs & Klarmann, 2012).

Replication. Replication occurs when the production of a product takes place in settings different from original product settings (Pritchard & Funk, 2010).

Assumptions, Limitations, and Delimitations

Assumptions

We assumed that participants for this study would give honest, thoughtful, and comprehensive responses to the questions. A further assumption was that participants would give accurate information on their experiences. Final assumprtion was that sample size for this study is adequate and represents the consumers of entertainment products throughout NYC. The research design and methodological procedures minimized the impact of these potential problems.

Limitations

This study of consumer behaviors that influences the purchase of replicate entertainment products has four limitations. First, the current study location was a limitation to a sample of convenience and results may not generally apply to other populations. Second, data collection for this study was limited to responses from questionnaires and triangulation of information using other means like field notes or followup questions was difficult. Third, participants may have relied on their memory and recollections and information they provided could be participants'perceived truth. Finally, information collected from participants lacked detail because questionnaire consisted of closed questions. Responses were fixed with less scope for respondents to supply answers which reflected true feelings on research topic. We addressed the limitation, through rich, thick description and analysis.

Delimitations

Delimitations include using a sample of consumer of entertainment products in NYC to collect data. Additional delimitation is the use of questionnaires consisting of closed ended question to collect data from up to 50 consumers of entertainment products about their understandings of product replication. Convenience samples of consumers of entertainment products in NYC who are 18 years and over may not representgeneralization to other locations.

Significance of the Study

Contribution to Business Practice

There is resultant increase in the practice of product replication, despite deterrent actions by authorities (Gopal & Gupta, 2010; Koster, 2012; Tade & Akinleye, 2012; Walls & Harvey, 2010). Wiedmann, Hennigs and Klarmann (2012) posited that entertainment managers may develop strategies to reduce the global appetite for replicate products with a better understanding of consumer purchasing behavior. An exploration of consumer purchasing behaviors may yield results useful to managers in deciding the appropriate business strategies required to inhibit product replication and reduce the negative effect on entertainment coporations revenue.

Results of this study might contribute to literatures on product replication by improving understanding of the phenomenon from entertainment consumer's perspective. As practice of replication continues to rise, entertainment leaders need to understand the strategies required to curb replication of products. The findings may lead to promotion of rights of owners, entertainment industry, and the society in general. Results of study may alsocontribute to the enhancement of a mission entertainment industry to champion intellectual property rights (Rosen, 2014), free and fair trade, and enrich peoples'lives through entertainment.

Implications for Social Change

Product replication affects goal of profit in the entertainment industry (Smallridge & Roberts, 2013). Curbing product replication is the responsibility of industry managers and stakeholders of entertainment corporations (Bonner & O'Higgins, 2010). Finding from this study may facilitate improvement of business strategies to inhibit product replication through understanding of consumer's perspective about the phenomenon. Inhibition of product replication may improve sales leading to a sustainable industry which may help entertainment leader's creat new jobs and sustain existing employments.

A Review of the Professional and Academic Literature

This literature review, consist of peer reviewed articles on entertainment product replication, effect of replication on product sales, and industry management strategies for inhibiting product replication. Additionally, literature for this study is sourced from the Business Source Premier; Science Direct; and ABI/INFORM Complete databases; Google Scholar; books; the Internet; trade journals; libraries; scholarly journals; and professional organizations.

We reviewed 187 peer-reviewed journal articles and books located by using key words such as *movie replication, replicated movie products, problems in the entertainment industry, internet replication, software replication, digital replication, hard goods replication, copyright replication, music replication, qualitative analysis in the entertainment industry, software copyrighting, qualitative research, quantitative research, the music industry, the movie industry, the film industry and loss of profit.* Other keywords include *quantitative analysis, managing entertainment and movie products, business strategies and entertainment, the effect of counterfeit product, analysis of counterfeit goods, demands for counterfeit products, intention to replication products, factors affecting replication behavior, case studies, research design, management strategies and entertainment replication, and effects of laws on product replication of entertainment products.*

The first section consists of software replication and represents the conclusions of relevant studies in chronological order. The second section was on music product replication and the findings of various past studies. The third section contains relevant studies of movie products replication and chronological presentation of conclusions of the studies. The fourth section is about digital replication and what researchers have learned. The fifth section describes the studies of Related Studies on product replication and the findings. The sixth section is on purchasing behaviors of entertainment consumers. The final section is the discussion of TPB as a conceptual framework and the application to research.

The digital age may have brought about an abundance of ways to copy or replicate original works (Rybina, 2011). The advent of Internet has resulted in an increasingly serious problem regarding the purposeful theft of intellectual property (Al-Rafee & Dashti, 2012; Wu, Chou, Hao-Ren & Mei-Hung, 2010). The practice of labeling infringement of exclusive rights in creative works as replication may be quite old. Typical entertainment-product replication methods may involve copying screened DVDs.

Replicate entertainment rips from original works were of poor quality due to noise from audience. Entertainment companies often release promotional copies of the film for critics and industry people to review in advance (Klinger, 2010). Replication may include removal of the promotional-copy-only message and releasing the film as a DVD rip. With advancements in technology, entertainment-product replication has grown at an alarming rate, prompting efforts by authorities to use copyright laws in curbing the problem (Rybina,

2011). Copyright laws have been set up for decades to ensure the intellectual property rights of creators, investors and stakeholders (Wu, et al., 2010).

Review of literatures reveals that a significant number of variables have not received adequate consideration from replication inquirers. Revenue loss to the practice of replication of entertainment products in 2005 was $1 billion in revenue, including film replication, Internet replication, bootlegged DVDs, and other forms of hard-good replication (Klinger, 2010). The actions of individuals who replicate entertainment products may be responsible for weak distribution and sales of entertainment products. Home video accounts for a portion of lost revenue and may be the market in which replication is most significant (Klinger, 2010). The interest in this study is to explore consumer behaviours which influence the purchasing of replicate entertainment products.

Software Replication

Software replication is the unauthorized transfer of technology; use or replication of intellectual property (IP) protected software (Chan & Lai, 2011). Chan and Lai argued that advancements in technologies resulted in the ineffectiveness of traditional instruments such as patents which managers used for protection of intellectual property. The production of unauthorized copies of software by individuals or businesses for resale or use in the workplace, school, or home, is an unconventional common (Chan & Lai, 2011).

Liao, Lin, and Liu (2010) used TPB to quantitatively examine perceived risk influence on attitude and intention toward using replicate software. Liao et al collected data through an online survey, offering the incentives of 500 New Taiwan dollar (NT$500, approximately US$15) cash prize to 305 participants in Taiwan. Liao et al. (2010) found that perceived prosecution risk affected intention to use replicate software, and perceived psychological risk was a reliable predictor of attitude toward using replicate software. Additionally, attitude and perceived behavior control contribute significantly to the intended use of replicate software. However, there was no support for a direct relationship between subjective norm and intention to use replicate software (Liao et al., 2010). Goode (2010) quantitatively examined the supply of replicate software for mobile devices against a backdrop of conventional desktop-replication theory using secondary data from 18,000 entries to a replicate software database. Goode (2010) found that a small number of replication groups accounted for the bulk of available replicate software and that methods for curbing copy protection for mobile devices were primary software based.

Woolley (2010) quantitatively examined whether the theory of reasoned action (TRA), adequately described music replication in the same fashion that TRA described software replication, using data from university students. Woolley found that the perception of friends' attitudes and behavior regarding replication determines students' replication behavior. Sheng, Shin, andChou (2010) qualitatively explored the intention to replicate software using their conceptual model. Sheng et al. (2010) concluded that perceived signal of replication emerges among users when they perceive a high percentage of software replicating individuals around them, ultimately leading to intention to imitate others' behaviors of replication, which then seem quite acceptable.

Rossman (2010) analyzed the music industry's anti-replication efforts and how filesharing is changing the music industry. Rossman found that surveillance and litigation efforts have driven some users from open peer-to-peer sharing to full secrecy. Phau and Ng (2010) quantitatively investigated the salient factors influencing consumers' attitudes and usage intentions toward replicate software with data from students and TPB. Phau and Ng (2010) found that neither informative nor normative influences are significant predictors of attitudes toward software replication.

Aleassa, Pearson, and McClurg (2011) quantitatively investigate software replication in Jordan using the theory of reasoned action and data collected from a sample of 323 undergraduate business students. Aleassa et al. (2011) found that attitudes toward software replication and subjective norms were significant predictors of intention to replicate. Aleassa et al suggested that ethical ideology, public self-consciousness, and lowself-control moderated the effect of variables of intention to replicate. Chan and Lai (2011) quantitatively considered Chinese computer users' ethical ideology and the relationship to software-replication attitudes and behaviors. Chan and Lai (2011) found that researchers divide Chinese computer users into situationist, absolutist, subjectivist, and exceptionalist ethical ideology types. When compared with situationists, absolutists, and exceptionalists, subjectivists have the least unfavorable attitude toward software replication and most frequently engaged in software replication (Chan & Lai 2011). Chan and Lai suggested that the government and authentic software developers and vendors focus on subjectivists as target audience for anti-software-replicate communications.

Dahlberg (2011) explored how the legal processes against Swedish file-sharing site, The Pirate Bay. Dahlberg focused on legal, cultural, and political relationships between activities and spatiality. McManus (2011) studied the emerging software industry in China. McManus (2011) found that China needed to overcome weaknesses in managerial and technical skills and focus on international markets, where it has strengths. McManus suggested that India offer some vital and practical lessons for China's emerging software industry.

Meissner (2011) discussed whether filesharing is an ethically problematic act. Meissner argued that one could not consider file sharing unethical, suggesting that it has positive and negative aspects, while proposing strategies to avoid filesharing with considerations surrounding other distribution alternatives. Meissner concluded that moral responsibility for file sharing lies with the film industry, not with the individuals who share files. Norazah et al. (2011) examined the factors that influence consumers' intention to purchase and use replicate software, using data from 289 consumers in Malaysia. Norazah et al found a significant and positive relationship exists between reciprocal fairness, procedural fairness, subjective norm, attitude, and consumer's intention toward software replication.

Peerayuth and Elkassabgi (2011) quantitatively examined the relationship between software replication and technological outputs in developed nations using data obtained from 28 industrialized countries between 2003 and 2007. Peerayuth and Elkassabgi found that firms in high-tech industries can benefit from software replication, when the level is limited to some optimal level while financially constrained. Small businesses may have more opportunity to use pirated software, to help increase technological productivity and innovation. Peerayuth and Elkassabgi concluded that in the longterm, high level of software replication will eventually hurt the competitiveness of a nation suggesting that software replication is a major problem that requires control in order to create the environment that encourages innovation. Lindgren and Linde (2012) investigated the possibility of recognizing parts of the collective act of online replication as a social movement. Lindgren and Linde collected data through qualitative interviews with young people and casual users in a case study of` Swedish participants. The authors found that online replication activities are not associated with political or legal dimensions. Online replication is an everyday culture builds on behavior taken for granted and made possible by technology (Lindgren & Linde 2012).

Andrés and Asongu (2013) examined governance mechanisms by which global obligations for the treatment of intellectual property rights (IPRs) are useful in the battle against replication. Andrés and Asongu used data collection from 2000 to 2010 for 11 African countries to conduct an empirical analysis and found that IPR laws are instrumental in tackling replication. The IPR is useful through government quality dynamics of rule of law, regulation quality, government effectiveness, corruption-control, and freedom of press. Andrés and Asongu concluded that there is need for a policy approach most conducive to expanding development and implementing an integrated system of both IPRs and corollary good governance policies. The authors suggested future research to confirm if those who copy or share software with others trust that softwares do not contain viruses, and if distributors of illegal copies trust the individuals not to report to the authorities (Andrés & Asongu, 2013).

Kigerl (2013) investigated the predictors of digital replication at the national level using data obtained from reports created by copyright industry representatives. Kigerl posited that poorer and less technologically advanced countries have a higher rate of replication, but lower absolute replication activity. Kigerl concludes that Software replication rates inversely relate to predictors such as population size, GDP, the number of Internet users per capita. The author suggested that policy ought to focus on wealthier nations, than the poorer nations, when targeting replication behavior.

Music Replication

Music replication is the practice of copying and distributing unlicensed copies of music or software productions (Woolley, 2010). The practice of downloading music without approval is a problem for the music industry (Tade & Akinleye, 2012). Legal loopholes through the Fair Use Act are contributory to problems of music replication. Under the Fair Use Act, a critic can quote from a work under consideration in the review (Cummings, 2010). Woolley (2010) explored the replication of music from original CDs and the downloading of music files from Internet.

Gunter, Higgins, and Gealt (2010) quantitatively examined the gap in researching diverse populations potentially involved in replication practice. The authors presented the first multivariate examination of engagement in digital music replication among middle school and high school students. Gunter et.al used data collection from a random sample of eight and 11th grade students in Delaware to predict involvement in music replication with demographics (sex, race, and class), educational achievement, and aspirations. Gunter et.al found that a smaller percentage of eight graders than 11th graders performed music replication. In addition, Gunter et.al found that students who earned bettergrades in school engaged in music replication, but were lesslikely to doso overall. The authors concluded that individuals spend more time performing other behaviors, such as replication, than focus on academic pursuits (Gunter, Higgins, & Gealt, 2010).

Jens and Mich (2010) examined the digitalization of music and its profound impact on the music industry, witnessed through conflicts in music industry, artists, consumers, as well as changes in business models. Jens and Mich found that adding even small uncertainty about the number of customers has serious implications. The authors showed that average profit per customer converges to zero for a monopolist selling individual units to customers, indicating a rather limited potential for traditional business model in the music industry. Jens and Mich determined that profits in music industry from selling individual copies of music files or CDs would continue to decline.

Hougaard and Tvede (2010) analyzed the market for digital music. The future business model of the music industry might be some combination of the eyeball model and open source model (Hougaard & Tvede, 2010). The authors concluded that future market form music would consist of online stores like those existing now but with considerably lower prices and lower profitability. Bonner and O'Higgins (2010) examined the issue of music replication under an ethical lens, with data collected from 84 students and workers through questionnaires on social-networking websites. They found that individuals unconventionally downloaded, despite viewing the act as immoral. Moreover, Bonner and O'Higgins found that some individuals' feel the act of replication is simply today's reality and nothing is wrong or immoral about illegal downloads. In addition, Bonner and O'Higgins found that those who illegally downloaded were less likely to admit the activity as being wrong. Active music fans were more likely to engage in undue downloading than passive ones. Based on previous research findings, Bonner and O'Higgins concluded that being a student versus having employment did not affect replication behavior (Hougaard & Tvede, 2010).

Seidenberg (2010) studied the new vehicles for delivering music in innovative ways. Seidenberg found that users connect directly with one another and companies like Napster, which supply peer-to-peer software, could not monitor, or control which data users share. Furthermore, Seidenberg (2010) found that there was no infringement, whether vicarious or contributory, under the Napster standard. Exum, Turner, and Hartman (2012) used quantitatively method to examine the diagnostic accuracy of self-reported intentions to offend (SRIO) scores by comparing participants' intentions to acquire illegal music files from a designated distributor to their actual attempts to obtainsuch data. Exum et.al, collected data from a convenience sample of 242 students from a southeastern university served as participants. Exum et.al (2012), found that approximately 7% of participants reported strong have definitive intentions to seek out the illegal files. The authors concluded that disconnect between what participants say in a research study, and what they do in the real world casts a shadow on the validity of self-reported intentions.

Hinduja (2012) examined the connection among music replication, general strain theory, and self-control theory, using data from a sample of university in the United States. Hinduja found that strain does not have a direct effect on music replication rather, self-control is a salient predictor. Turri, Smith, and Kemp (2013) quantitatively examined how emotional, or affect-based brand relationships, are developed in online social communities. The authors used data obtained from Internet-based survey administered to 422 students at a school in the south-western part of the United States. Turri et al found that affectively committed users were less tolerant of and less willing to engage in copyright theft. Turri et al concluded that an affective tie to the branded artist result in consumers expressing greater support for musician's artistic vision leading to greater brand advocacy from consumers.

Movie Replication

Movie replication occurs when individuals who do not own rights to original work (Banutu-Gomez, 2013) replicate original movies. Reducing replication rate and preventing counterfeit may shed light on several significant economic benefits for the market or economy. Neenu and Lobo (2011) posited that the film industry relied on its box-office collections, sale of Blu-ray discs, DVDs, and video CDs to create profit. Individuals who replicate movies seriously threaten the profits.

Walls and Harvey (2010) examined the surviving brick-and-mortar market for replicate DVDs in Hong Kong. Walls and Harvey studied the flow of customers, the logistics of marketexchange, and frequency of market disruption due to enforcement of law. The authors obtained a sample of 30 DVD movies from a brick-and-mortar market for replicate DVDs. Ho and Weinberg (2011) examined how different segments in the movie market respond to three marketing drivers namely, prices, product availability, and viewing channels (including replication). The authors collected data from the questionnaire administered in three shopping malls of a major Canadian city. The need to segment consumers of pirated products by channels of acquisition was suggested by individual users to portray that replication as immoral, unethical and has limited impact (Ho & Weinberg).

Koster (2012) explored the fight by French authorities against Internet replication with Hadopi Law. Koster concluded that three primary objectives of government with Hadopi include enforcing copyright law on the Internet through legal actions against violators. Other objectives are to educate Internet users about illegal versus legal activities with respect to copyright law, and facilitate development of Internet services providing legal access to copyrighted works. Some modest positive changes in behavior of French Internet users resulted from the use of Hadopi law (Koster, 2012). Dolinski (2012) explored the Legal Boundaries between Internet replication and Legal Exchange of Files through Internet. The author determined that Corporations intensified fight against Internet replication through a series of free educational programs shaming individuals who engage in replication practice as well as lobbied for appropriate laws. In the past, development of individual legislative acts was dedicated to fighting Internet replication which raised some salient questions in digital world.

Wing (2012) explored the debate over copyright replication and its control in Western countries, particularly the use of graduated response laws in countries like the UK. Wing found that legal responses so far have proved to be ineffective due to some identifiable problems and suggested that legal responses may continue to fail. Wing concluded that it is time to take a different more pragmatic and coordinated approach to the on-line replication. Banutu-Gomez (2013) explored effects of replication and counterfeiting on international economy. The author found that individuals who replicate can counterfeit virtually everything and counterfeiting went beyond CDs in all economies. A primary reason product replication continues to occur is the ease which individuals can export replicate and counterfeited components (Banutu-Gomez, 2013). The author suggested that though most governments have laws against product replication, enforcement has not been strong enough.

Digital Replication

Digital replication is the act of copying digital goods, computer software, electronic documents, and digital media without explicit permission of copyright holders or consumption of illegal copies of digital services (Taylor, 2012). The use of Internet may facilitate this form of replication by allowing the act to take place anonymously. The practice of digital replication affects various stakeholders, like recorded music, movie and software industries, distributors and consumers (Vida, Mateja, Kukar-Kinney, & Penz, 2012). Individual who replicate can download products without cost or licensing. Digital replication continues to perplex service marketers who produce replicable digital products such as music, movies, software, etc (Taylor, 2012). Efforts to curtail replications involve promoting public awareness through consumer education and protection of intellectual property rights through legal threats and actions against operators facilitating digital replication activities (Vidal et.al, 2012).

Morris and Higgins (2010) studied digital replication and social learning theory using data collected from 585 college students in different universities. The authors found that individuals are likely to follow social-learning process, which leads them to digital replication and modestly supports social-learning theory concerning digital replication. Lorde, Devonish, and Beckles (2010) examined factors that affect propensity for digital replication in Barbados with data from 390 Barbadian participants. Lorde et al. found that attitude and beliefs about replication, serve to predict the digital replication intentions of Barbadians. Individuals with higher levels of education tend to have a higher tendency to replicate due to the influence of peers (Lorde et al., 2010).

Danaher, Dhanasobhon, Smith, and Telang (2010) examined the impact of digital distribution on sales and Internet replication with data from two large data sets from Mininova and Amazon.com documenting levels of replication and DVD sales for NBC and other major networks' contents. Danaher et al. found that replication increase for drama programming was insignificant, whereas increase for action movies was significant at approximately 11%. Danaher et al. determined that comedy and science fiction replication increased considerably, above 20%. Danaher et al. (2010) concluded that young men disproportionately replicate which aligns with conventional wisdom that action, comedy, and science fiction appeal to young men.

Williams, Nicholas, and Rowlands (2010) evaluated the literature on digital consumer behavior and attitudes toward digital replication. The authors found that for studies are mainly concerned with behaviors and attitudes of young people with minimal studies looking at demographic differences. The authors found, that social and situational factors affect likelihood to replicate more than ethical considerations. Anonymity is a strong indicator replication practice, whereas literature is ambiguous about whether punishment acts is a deterrent to replication practice (Williams et al., 2010). Yoon (2010) studied ethics theory in digital replication using data from 270 Chinese university students. Yoon found that moral obligation and justice is derived from attitude, subjective norms, and perceived behavioral control. The authors found that perceived benefits; perceived risks and habits also affect attitudes of individuals toward digital replication.

Yu (2010) focused on a direct comparison between the propensities for committing digital replication and that for stealing. Yu collected data from 501 collegestudents from six universities in the United States. Yu found that stealing and digital replication propensities are two separate concepts. The author suggested that digital replication propensity is not a direct result of low morality, whereas stealing propensity is strongly associated with low morality. According to Yu, age plays a role in digital-replication propensity, whereas in stealing propensity, age is not a factor. Digital replication propensity is more likely to result from the tendency to explain digital replication using neutralization techniques.

Downing (2011) used cyber ethnography to study the community of retrograde gamers regarding digital replication of retrograde and contemporary video games. Downing (2011) found that sub cultural demands created the conditions under which other goals predominated over consumer desires on a cultural level. Rybina (2011) examined influence of social norms and standards, perceived risk of physical injury, and consumer expertise in filesharing on digital replication in post-Soviet economies. Rybina used a sample of 226 respondents from Russia, Kazakhstan, and Kyrgyzstan to examine factors influencing intentions to replicate digital music. Rybina found that the higher consumer perception that digital music replicating is socially acceptable, the higher probability of digital-music replicating in post-Soviet countries. Rybina (2011) found no significan trelationship between perceive drisk of personal harm and intention to replicate music in post Soviet countries. Rybina also found that increases in consumers' expertise of music filesharing increased the probability of digital replication of music.

Panas and Ninni (2011) examined the influence of attitude, subjective norms, perceived behavioral control, moral responsibility, perceived equitable relationship, and deterrence effect of the legislation in ethical decision making about electronic replication. Panas and Ninni used data collected from 799 undergraduate students of Athens University. The authors suggested that men and women differed in the manner in ways Internet communication influences the intention to download music inequitably. Panas and Ninni found that for men, the most significant variables associated with intention toward music replication were attitude, perceived behavioral control, subjective norm, and moral obligation; whereas for women, attitude, subjective norm, perceived behavioral control, and perceived equitable relationship were most significant.

Oguer (2011) considered the decision to enforce the Hadopi Act rather than support the global licensing of replication networks. Oguer found that although countries, such as the United States and the UK, have chosen prosecution against replication, some like Brazil considered the choice of global license, whereas, others such as Belgium, studied both options. Taylor (2012) studied how well digital replication self-report intentions predict actual digital replication behaviors in marketing research. Taylor collected data from 297 students of two large-section Marketing Principles classes at a Midwestern university in the United States. Taylor determined that there is a strong predictive relationship between digital replication intention formation and subsequent actions.

Vida, Mateja, Kukar-Kinney and Penz (2012) investigated consumer perceptions of own risk and benefits of digital replication behavior as determinants of the justification for such action and the consequent future replication intention using a mixed method. Vida et al, collected data with a sample comprised of 1,213 respondents from six nations. The authors found that perceived technical risk is less influential on cognitive processes, but have a negative impact on future replication intent. Blankfield and Stevenson

(2012) explored the technological methods which United Kingdom and United States publishers and their representative bodies are using to tackle the growing challenge of e-book replication. Blankfield and Stevenson collected data through interviews with industry experts, consultants, administrators, and representatives. The authors brought together knowledge previously published and considered it separately using systems that include Publishers Association Copyright Infringement Portal, and Digital Rights Management in its variousforms.

Larsson, Svensson, de Kaminski, Rönkkö, and Olsson (2012) explored replication using secondary data collected from 75,901 respondents overall in the study through BitTorrent tracker the Pirate Bay website. Larsson et al, found that online anonymity practices in the filesharingcommunity are dependent on legal and social norms (i.e more enforcement meant more anonymity). Online anonymity in the file-sharing community is an active counter measure against legal action (Larsson et al., 2012). Al-Rafee and Dashti (2012) studied a behavioral model in the United States and the Middle East using TPB in the context of digital replication. The authors collected data from 613 (328 Kuwaitis, 285 U.S. citizens) college undergraduate students. Al-Rafee and Dashti found that the attitude was significant in the Middle East, compared to the United States.

Related Studies

Ethics is essential for personal development of future business leaders while ethical values form the fundamentals of ethical culture within organizations and business environments, which operates on self-regulation (Nga & Lum, 2013). Nga and Lum (2013) explored the influence of gender, and course majors on unethical behavior intentions among Generation Y under graduates. The authors used a sample of 245 undergraduates from a private higher education institution in Malaysia and theory of planned behavior. The authors posited that egoism, utilitarianism and magnitude of consequences, exert significant influence on unethical behavior intentions. In terms of gender, the reason for unethical intentions among males included egoism and peer affect while influence among females included utilitarianism and magnitude of consequences (Nga & Lum, 2013). Nga and Lum suggested an extension of research to actual behaviors and adult samples while accounting for religiosity and age.

Consumers' attitudes, moral intensity, and perceived risk influence intention to buy while moral intensity and perceived risk determine attitudes (Koklic, 2011). Unfavorable attitude toward knowingly purchasing replicated products along with perceived risk significantly shaped individuals intention to engage in purchasing replicated products (Koklic, 2011). The underlying determinants of unfavorable attitude toward knowingly purchasing replicated products are perceived consequences for society, and perceived consequences for individuals (Koklic, 2011).

Individual characteristics may influence propensity toward downloading and deterrent messages communicating that downloading is illegal or comparing downloading to stealing are unlikely to deter downloading behavior (Robertson, McNeill, Green & Roberts, 2012). Individuals who downloaded showed little concern for the law; therefore, illegally downloading music through peer-to peer networks has persisted in spite of legal action to deter the behavior (Robertson et al., 2012). According to the authors, Men and women were equally likely to download and the factors characterizing downloading were similar for men and women. Robertson et al., found that downloaders have less ethical concern, engaged in other illegal behavior, and indicated a propensity to steal a CD if downloader's know there is no risk of been caught.

Jančić (2010) explored a set of complex legislative proposals to reform the telecommunications sector of the European Union to find out whether an administrative organ or only a judicial body could impose such a prohibition. The author concluded that action is a political process that crosses levels, not institutional incarnations, and that European politics mirrored of various national politics. Meyer and Leo (2010) studied the warning and sanction mechanisms aimed at fighting online replication. Meyer and Leo analyzed the graduated response debate in the European Union, and the current initiatives in France and the United Kingdom. Meyer and Leo concluded that a graduated response was a contested means of deterring online copyright infringement.

Pang (2010) explored Hong Kong Cinema and the use of dialect in the mass media particularly in China. Pang concluded that cinematic autonomy of the Chinese province has a historical precedent. The province implements its own regional film and censorship policies. Burke (2010) discussed the issue of online replication, counterfeiting, and the role played by the United States in global efforts to fight the problem. Burke found that intellectual property theft costs the U.S. economy more than $100 billion annually and concluded that traders worldwide legitimately sell as minimal as one in three music CDs and one in 20 musical downloads.

Watson and Dow (2010) examined auditing compliance. The authors determined that most shareholder value was lost due to strategic, operational, and business risks, rather than financial risks. Internal auditing was shifting its focus from financial compliance auditing to nonfinancial concerns. The authors provided an active-learning opportunity for students to assume the role of an internal auditor conducting a compliance audit to help students gain the necessary skills to conduct audits. Watson and Dow (2010) suggested that Sarbanes-Oxley Act of 2002 shifted the focus from internal auditing (). McLennan and Le (2011) explored the relationship between intellectual property rights and the growth rate of per capita GDP during 1996–2006 periods in 71 countries using software replication data as a proxy for intellectual property-rights infringements. The authors found that countries with increasing rates of property rights violations tended to have lower growth rates, whereas countries with strong governance to enact policies that protect property rights exhibit increasing growth.

Economic and social strain events are conditions that individuals dislike (Lin & Mieczkowski, 2011; Carlo, Padilla-Walker & Day, 2011) found that economic pressures can causestrain. Failure to achieve positive valued goals like education, good accommodations, transportation, and valued work comprises the economic strain (Lin & Mieczkowski, 2011). Strain also involves removal of positively valued stimuli, such as losing one's culture, a good friend, or a relative (Lin & Mieczkowski, 2011). In addition, strain is the subjection to negatively valued stimuli such as receiving negative treatment from associates, or exposure to strange cultures (Lin & Mieczkowski, 2011).

Consumption

Consumption is the use of goods and services with exchangeable value by individuals (Grauerholz & Bubriski-McKenzie, 2012). With regard to consumption values for choosing to purchase or stay away from any product, various values influence consumers' purchase choices. The consumer's decision-making process is personal perceptions and non-personal perceptions like perceived conspicuousness, perceived uniqueness and perceived quality. Norazah, Ramayah, and Norbayah (2011) suggested that increase in entertainment product replication is a result of a mixture of socioeconomic, cultural, political, and technological issues. Andersen and Frenz (2010) suggested that there are minimal qualitative researches on replication of entertainment products.

A business manager's key to success in a highly competitive and saturated market, is the knowledge of consumer's consumption patterns. To develop attractive offer of products and satisfy customer's needs, business managers need to recognize and understand factors which influence purchases (Kurajdova & Taborecka-Petrovicova, 2015). In this study, we focused on non-deceptive replicate products, which consumers know with certainty, are not original products, and affect the price, market share and profitability of original products (Zhang, Hong & Zhang, 2012).

Consumer Behavior

A consumer is the person who buys or uses goods or services (Patwardhan, Flora & Gupta, 2010). Consumer behavior refers to the selection, purchase and consumption of goods and services for the satisfaction of their wants (Latuszynska, Furaiji & Wawrzyniak, 2012). Everyone has certain basic human needs that serve as motivation for him or her to take action, including buying action. According to Chinaka (2016), the central idea behind purchasing a product or service is to satisfy needs and wants. Economic wants are desires that individuals can satisfy by consuming a good, service or leisure activity (Morrow, 2015). It is imposible for anyone to satify all wants because resources are limited and as a result, individuals make choices to satisfy some wants and give up others.

Consumers possess specific belief and attitude towards a variety of products. Since such beliefs and attitudes affect consumer buying behavior which should interest business managers. According to Parnell (2011) and Porter (1980), business managers can pursue superior performance by either establishing a cost-leadership position or differentiating its products and services from those of its competitors. In addition, entertainment business managers can change beliefs and attitudes of customers by launching special campaigns in this regard.

Factors Affecting Consumer Behavior

The initial process of consumer behavior involves efforts to find what commodities the consumer wants. Thereafter, the consumer should select commodities with promises of superior benefits. After selecting commodities, consumers may estimate the available money. Finally, the consumer analyzes prevailing prices of commodities and decides on the commodities to consume. Other factors which influence purchases by consumers include social, cultural, personal and psychological (Rani, 2014).

1. Cultural Factors
Consumer behavior is influenced by cultural factors such as: buyer culture, subculture, and social class (Dave & Patel, 2016).

Culture

Culture is a part of every society and a significant cause of an individual's wants and behavior. Impact of culture on buying behavior differs from localities to localities and managers need to be very careful in analyzing the culture of different groups.

Subculture

Each culture consists of diverse subcultures such as religions, racial groups, nationalities, etc. Managers can use these groups by segmenting the market into diverse small portions.

Social Class

Every society consists of social classes which is important to business managers because individuals in a given social class have similar buying behavior. Business managers may to specifically produce to satisfying different social classes. Income is not the only determinant of social classes. Other factors of social class include wealth, education, occupation etc.

2. Social Factors
Social factors also influence the buying behavior of consumers. Significant social factors include reference groups, family, role and status (Uchenna, 2015).

Reference Groups

Reference groups have potential in forming a person attitude or behavior. The impact of reference groups varies across products and brands. According to Uchenna (2015), influence of reference groups on buying behavior of consumers for tangible products such as dress, shoes, car etc is high. Reference groups includes individuals who influences others because they posses special skill, knowledge or other characteristics.

Family

Family members strongly influence the buyer behavior of consumers (Rani, 2014). Therefore part of a business manager's job is to strategize on the roles and influence of consumer's family. However, buying roles change with changes in consumer lifestyles.

Roles and Status

Individuals function in different roles and status of their societies based on the groups, family, and organization etc. to which they belong. An individual could work in an organization as sales consultant and also have another role as a mother. Such individual's buying decisions may be influenced by the role and status.

3. Personal Factors

Personal factors may affect the consumer behavior. Some personal factors that influence consumer buying behaviors include lifestyle, economic situation, occupation, age, personality and self-concept (Latuszynska, Furaiji & Wawrzyniak, 2012).

Age

Consumers tend to change their purchases of goods and services with the passage of time. Therefore, age and life-cycle have potential impact on the consumer buying behavior. Family life-cycle consists of different stages such as teen, young, old, single, married, unmarried etc. The age grouping is useful to business managers for developing appropriate products for each segment.

Occupation

The occupation of an individual significantly influences their buying behavior. Indivuduals who work in coporate positions of an organization purchase coporate foothwares while others like mechanics and low level workers in the same organization will purchase sturdy shoes.

Economic Situation

Consumer economic situation has great influence on buying behavior. If the income and savings of a consumer is high that consumer will purchase more expensive products. On the other hand, an individualwith low income and savings will purchase inexpensive products.

Lifestyle

Lifestyle is the way an indididual live in a society expressed by the things in their surroundings. Lifestyle of customers influences a wide range of consumer behavior, such as choosing a vacation destination or choosing activities while at the destination (Naylor & Susan, 2002). Lifestyle is determined by a consumer's interests, opinions, and activities which shape their pattern of acting and interacting in the world.

Personality

Personality influences how individuals think, attend, learn, feel, and act in social contexts particularly in consumption settings (Tao, 2013; Wells, Burnett, & Moriarty, 2005). Personality changes from person to person, time to time and place to place. Personality is a totality of the behavior of an individual in different circumstances. It has different characteristics such as: dominance, aggressiveness, self-confidence etc which can be useful to determine the consumer behavior for particular product or service.

4. Psychological Factors

According to Sharma (2014), modern shoppers buy things, to satisfy psychological needs that make them feel good. There are four important psychological factors affecting the consumer buying behavior. These are: perception, motivation, learning, beliefs and attitudes.

Motivation

Motivation is the internal force that directs behavior toward fulfillment of needs (Tokuyama & Greenwell, 2011). Motivational factors strongly influence consumers' decision-making processes (Cohen & Avrahami, 2005). Every person has different needs such as physiological needs, biological needs, social needs etc. The nature of the needs is that, some of them are most pressing while others are least pressing. Therefore a need becomes a motive when it is more pressing to influence an individual seeking satisfaction.

Perception

Perception is the process through which an individual select, organizes and interprets information in order to make sense of it (Chinaka, 2016). Chinaka suggested three different perceptual processes which are selective attention, selective distortion and selective retention. In case of selective attention, marketers try to attract the customer attention. Whereas, in case of selective distortion, customers try to interpret the information in a way that will support what the customers already believe. Similarly, in case of selective retention, marketers try to retain information that supports their beliefs.

Beliefs and Attitudes

Consumers possess specific belief and attitude towards various products which make up product image and affect consumer purchasing behavior (Chinaka, 2016). Consumer beliefs and attitudes should interest entertainment business managers because they make up brand image and affect consumer buying behavior. Managers can address or change the beliefs and attitudes of customers by launching special campaigns focused on consumer beliefs and attitudes.

5. Marketing mix

Marketing mix is the set of selling tools for helping company managers aim target customers in marketing (Hu, 2011; Shah, 2014). Consumers are exposed to stimuli that are identified with the four P's: product, price, place and promotion (Blythe, 2008; Latuszynska et al., 2012).

Promotion

Managers use promotion to help increase awareness among consumers. Consumers come to know about offers only through promotion (Shah, 2014). Promotion represents all of the communications that a marketer may use in the marketplace.

Promotion distinct elements are advertising, public relations, personal selling and sales promotion.

Place

A way of getting the product to the consumer and/or how easily accessible it is to consumers. Place is the availability of product to targeted customers (Shah, 2014). Business managers may offer multiple channels to reach to consumers.

Products

Product is a good or service offered by company to consumers in the market (Shah, 2014). Product is offered for attention, acquisition or consumption, to satisfy the needs and wants of customers. If the product can be differentiated from competitor it will give unique value to customers (Shah, 2014).

Price

Price is what the customer will pay for the purchase of product (Shah, 2014). Price mix is very important factor that influence consumer's purchase decision making.

Costs and Benefit

Benefits are the change in individual well-being which policy induces while costs are generally measured in terms of monetary costs of resources (Wolfson, 2001). Business buyers usually want to see a cost-benefit analysis. Cost-benefit analysis refers to a narrower class of procedures of evaluation in terms of the net benefits to individuals (Wolfson, 2001). The concept of costs and benefits encompass an area of economics which relates to rational expectations and rational choices. In line with the concept of costs and benefits, individuals are likely to make the choice which has the most benefit to them, with the least cost under any giving situation. The principle of costs and benefit extends beyond financial transactions and relate to choice that provides more in benefits than it costs.

In line with the principle of Cost and benefit, a beer consumer will buy the best beer he or she can afford, not, neccessarily, the best tasting beer in the store. In order to explain individual behaviour, economic science uses the rational choice model. The main assumption in the rational choice model is that the behaviour of all economic actors is perfectly rational (Mcclennen, 2010). The perfect form of rationality is based on the principle of maximization that consumers always tend to maximize utility, while manufacturers tend to maximize profit.

Rational Choice Theory

Foundation of Rational Choice Theory (RCT) is the basic insight proposed by Joseph Schumpeter in early 1960's that public forum can be modeled on a market (Maloy, 2008). According to Maloy (2008) rational choice theorists have been constructing their models on the premise that political is fundamentally similar to economic behavior proposal by Joseph Schumpeter. Researchers use the RCT to attempts to explain all conforming and deviant social phenomenon in terms of how self-interested individuals make choices under influence of their preferences (Krstic & Krstic, 2015).

The RCT treats social exchange as similar to economic exchange where all parties try to maximize their advantage or gain, and to minimize their disadvantage or loss. Rational choice theories rest on a central premise that individuals behave in ways that maximize their rewards or benefits and maximize utility. In order to make choices that maximize benefits, rational thinking is used to evaluate the choices available. For each choice or option, three elements are used to calculate the resulting net benefits. First, there is the actual value of benefits to be reaped as the result of that choice. Second, an individual considers the costs associated with that choice. Third, one must take into account the forgone costs of that choice - that is, the benefits that could have been obtained if one made an alternate choice. Thus, rational choice theory assumes that choices are intentional, conscious, and rational (cite).

The basic idea behind RCT' are that individuals base their behavior on rational calculations, and act with rationality when making choices. In addition, the choices which individuals make are aimed at optimization of their pleasure or profit. This concept is in line with applications in economics and marketing, and in criminology and international relations. The key problem with the rationality assumption is that explanatory power of rational choice theory is significantly reduced in specific cases because it is so general that it can explain everything (Krstic & Krstic, 2015).

Theory of Reasoned Action

The Theory of Reasoned Action (TRA) was developed in 1975 by Dr. Icek Ajzen and Dr. Martin Fishbein to study human behavior for developing appropriate interventions. The TRA was designed to make statistical generalizations predicting individuals behavior and posits that intentions must precede consumer behavior (Ajzen & Fishbein, 1975). In line with the TRA, individuals make conscious choices based on two factors;

a.) how strongly they perceive the benefits to lead to a positive outcome, and
b.) the social norms, risks, and rewards they associate with that choice.

Researchers use TRA to predict the attitudes and behaviors of large groups of people. Orr, Thrush and Plaut (2013) argued that TRA basically discusses how people decide to perform a certain behavior. The TRA reasons that people consider their actions before they decide to perform or not perform certain behavior (see Figure 1.0). Intention is a major component of this theory. The TRA assumes that individuals will usually act upon their intentions. The specific intentions are comprised of two major attributes: an individual's attitude toward a behavior, basically whether it is right or wrong;

and an individual's beliefs regarding social pressures to either perform or not perform the behavior.

An assumption of TRA is that individuals are usually quite rational and make systematic use of information available to them. The concept of TRA is useful for predicting and understanding behavior and attitudes. People consider the implications of their actions before they decide to engage or not engage in a given behavior. The TRA is a framework, which researchers may use to look at behavioral intentions rather than attitudes as the main predictors of behaviors.

A weakness of TRA is the self-reporting technique which researcher use to learn respondents' attitudes and subjective norms. Self-reported data tends to be highly subjective and less valid (Coleman, Bahnan, Kelkar & Curry, 2011). Another limitation stems from the assumption that all behavior is voluntary and consciously analyzed before hand. In addition, the TRA fails to predict or explain irrational or impulsive behaviors. Coleman et al. (2011) posited that behavioral intention does not always lead to actual behavior when an individual's control over the behavior is incomplete.

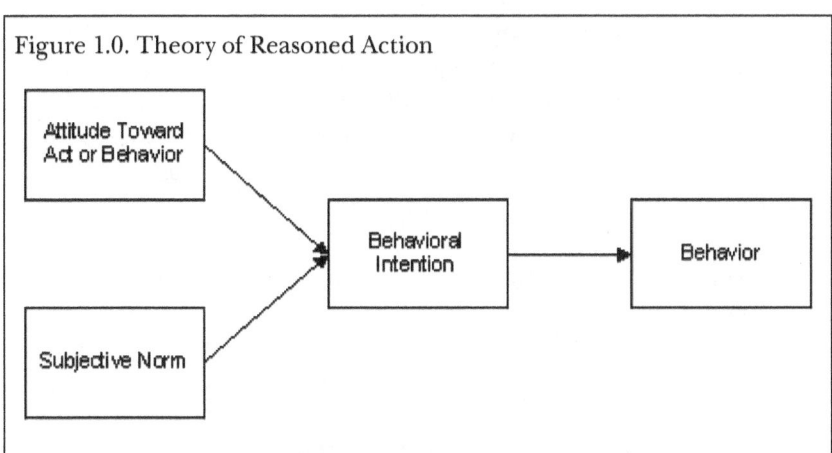

Figure 1.0. Theory of Reasoned Action

Theory of Planned Behavior

To address the issue TRA's failure to predict or explain irrational or impulsive behaviors, Ajzen (1991) propose the Theory of Planned Behavior (TPB), which extends the TRA by adding the perceived behavioral control component to account for behaviors that occur without a person's volitional control (see Figure 2.0). According to Coleman et al., (2011), TPB outperforms

TRA in predicting favorable behavioral intentions (FBIs). Theory of Planned Behavior (TPB) was added to the existing model of reasoned action to address the inadequacies that Ajzen and Fishbein had identified through their research using the TRA.

As the TRA began to take hold in social science, Ajzen and other researcher realized that TRA was not adequate and had several limitations (Godin & Kok, 1996). A significant limitation of the TRA was with people who have little or feel they have little power over their behaviors and attitudes. Ajzen observed that aspects of behavior and attitudes as being on a continuum from one of little control to one of great control. To balance these observations, Ajzen added a third element to the original theory which he called the concept of perceived behavioral control. The addition of this element resulted in the newer theory known as TPB.

Purpose of TPB:

- To predict and understand motivational influences on behavior that is not under an individual's volitional control.
- To identify how and where to target strategies for changing behavior.
- To explain virtually any human behavior such as why a person buys a new car, votes against a certain candidate, is absent from work or engages in premarital sexual intercourse.

Assumptions

1. Human beings are rational and make systematic use of information available to them.
2. People consider the implications of their actions before they decide to engage or not engage in certain behaviors.

If a person perceives that the outcome from performing a behavior is positive, that person will have a positive attitude forward performing that behavior. The opposite can also be stated if the behavior is thought to be negative. If relevant other individuals see performing a person's behavior as positive and the person is motivated to meet the exceptions of the relevant individuals, then a positive subjective norm is expected. If relevant other individuals see a behavior as negative and an individual wants to meet the expectations of these other individuals, then the experience is likely to be a negative subjective norm for the individual. Attitudes and subjective norm are measured on scales like the Likert Scale using phrases or terms such as like, unlike, good, bad, and agree' disagree. The intent to perform a behavior depends upon the product of measures of attitude and subjective norm. Glanz, Lewis & Rimer, (1997) suggested that a positive product indicates behavioral intent.

The TRA works most successfully when applied to behaviors that are under a person's volitional control. If behaviors are not fully under volitional control, even though a person may be highly motivated by own attitudes and subjective norm, that person may not actually perform behavior due to intervening environmental conditions. The TPB was developed to predict behaviors in which individuals have incomplete volitional control.

The major difference between TRA and TPB is the addition of a third determinant of behavioral intention, perceived behavioral control. Perceived Behavioral control is determined by two factors; Control Beliefs and Perceived Power. Perceived behavioral control indicates that a person's motivation is influenced by how difficult the behaviors are perceived to be, as well as the perception of how successfully the individual can, or cannot, perform the activity. If a person holds strong control beliefs about the existence of factors that will facilitate a behavior, then the individual will have high perceived control over a behavior. Conversely, a person will have a low perception of control if the person holds strong control beliefs that impede a behavior. This perception can reflect past experiences, anticipation of upcoming circumstances, and attitudes of the influential norms that surround an individual (Mackenzie & Jurs, 1993).

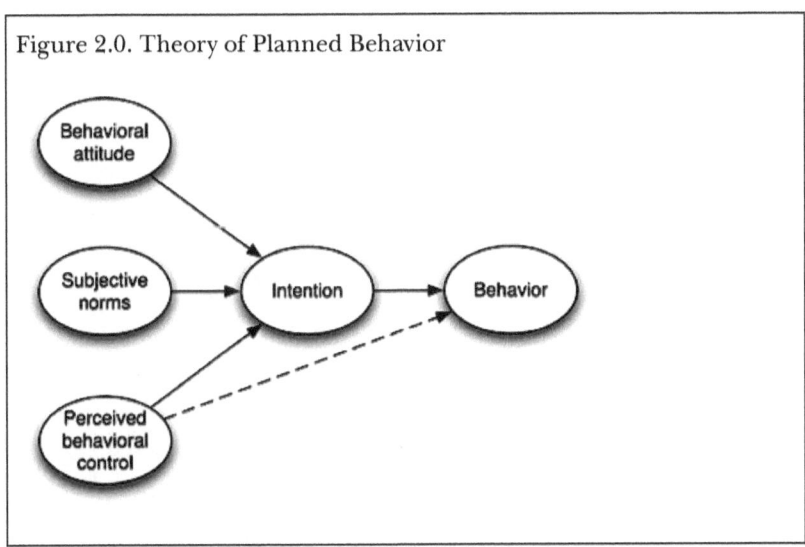

Figure 2.0. Theory of Planned Behavior

Behavioral attitude

Subjective norms

Perceived behavioral control

Intention

Behavior

Transition and Summary

Section 1 contained discussions on the background of the study, the problem statement, purpose statement, nature of the study, research question, definition of terms, and conceptual framework. In addition, section 1 included discussion on assumptions, limitations, delimitations, significance of study, reduction of gaps, and implications for social change. Researchers who have conducted relevant researches on entertainment product replication includes, (Al-Rafee & Dashti, 2012; Klinger, 2010; Wu, Chou, Hao-Ren & Mei-Hung, 2010; Chan & Lai, 2011; Liao, Lin, & Liu, 2010; Andrés & Asongu, 2013). Understanding the entertainment consumers purchasing behaviors might result to formulation of strategies required by managers to inhibit replication of entertainment products leading to viable entertainment corporations and saving jobs.

Section 2 contains a discussion on the role of individuals performing research, participants, research method and design, population and sampling, reliability and validity. In addition, section 2 contains details relating to data collection, including (a) instruments, (b) data collection techniques, (c) data organization techniques, and (d) data analysis techniques. Finally, we identified and described the research instrument for this study.

SECTION 2

The Project

Product replication is the unwarranted practice of copying products, without explicit permission from and compensation to the copyright holder (Marcum et al., 2011). Simburg et al. (2012) suggested that replication of entertainment-product constitutes infringement on the Pro-IP Act of 2008. In addition, practice of entertainment product replication impacts on unemployment situation in the United States (Al-Rafee & Dashti, 2012). Entertainment-product replication is a growing practice that particularly affects NYC (N. Akaeze, 2016). Product replication may continue because of advancement in technology and icreased Internet usage if nothing is done to inhibits the practice. We were interested in exploring the consumer behaviors which influence complaisance towards purchasing replicate entertainment products in NYC.

The audiences for this study include Entertainment industry producers, actors, distributors, and academics. Product replications impact on entertainment corporation manager's ability to manage sales and profits effectively (Bonner & O'Higgins, 2010). Section 2 includes (a) purpose of research, (b) role of researcher, (c) participants, (d) research method, (e) research design, (f) population and purposeful sampling. In addition section 2 consists, Sampling, selection criteria, and ideal numbers, (h) measures taken to ensure ethical research, (i) data collection, data collection instrument, organization techniques, (j) data-analysis techniques, and (k) reliability and validity.

Purpose Statement

The purpose of this qualitative multiple case study is to explore consumer behaviors which influence complaisance towards purchasing of replicate entertainment products. We designed questionnaire to collect data from up to 50 consumers of entertainment products who purchased music and movie products for atleast two years using convinient sample of participants in NYC.

Findings may assist entertainment managers who gather the ancillary talents, produce and package resultant products (Fradley, 2010; Oliver, 2010) to curb products replication, improve sales and increase profit margins. In addition, business impact of this study may include prevention of loss to replication of products and improvement to corporate profit. The social impact of this study may include sustainable and viable entertainment industry that may save jobs in the entertainment industry and affiliate businesses.

Role of the Researcher

The authors share between them a total of more than two decades experiences in consultancy services, marketing management and human resources services in the United States and globally. Industry experiences include banking and financial services, Auto sales management, and petroleum marketing management. As an auto sales manager, Dr. Chris was responsible for decision making and for overseeing and managing customer satisfaction towards products and service on a daily basis, and for competitive advantage. Dr. Nana was an HR generalist who specialized in consulting and training management and staffs towards product sales and improvement. Our roles in this study included sampling of participants, design and setting up of research questionnaires on Monkey survey website, inviting participants to contribute to research through social media, e-mail, and collecting data for analysis.

Qualitative researchers may become personally involved in the phenomenon under study (C. Akaez, 2016). We identified assumptions, values, and biases early in the study. In addition, we provided statements of ethical disclosure through the questionnaires to participants before accessing responses to research questions. We provided e-mail accesses to participants for answers and clarification about research questionnaires. However, the use of online survey instruments reduced personal interactions with participants and potential biased responses. There was adequate precautions to reduce biases during the data collection process, data organization, and data analysis. We made every effort to verify that data collection is reliable, and data interpretation is correct without bias. Upon completion of study, we provided detailed and accurate summary of findings. Researchers have the responsibility to present findings, unbiased and honestly, to enable readers understand and encourage further research (N. Akaeze, 2016).

Participants

Vulnerable persons are those who are unable to protect their self-interests because they may have inadequate control, reasoning, education, support, stability, or other essential features to defend their self-interests (Gwyn & Colin, 2010). A group that may be vulnerable for this study is non-English speaking entertainment product consumers that may lack comprehension of the questions because they lacked proficiency of English language. To establish trust and enhance confidentialty, participants were not required to identify their names or any personal information during the study. In this study, focus was on non-deceptive replicate entertainment products, which consumers know are not original products.

Participants were informed that replicate entertainment products are unauthorized, reproduced copies entertainment products. Questionnaire was the instrument of choice because only a few people want to identify with unconventional behaviors considered to be unacceptable (Yu-Chin & Hsu, 2013). Nulty (2008) suggested that online questionnaires yield up to 30% response while paper questionnaires of the same participants yield about 60% response. Therefore, to get up to 50 responses, 180 questionnaires targeting consumers of entertainment products were sent out. We used convinient sampling strategy to select participants who must be over 18years of age and have atleast two years of experience purchasing entertainment products in NYC. Because of the simplicity and ease of research, convenient sampling facilitated cost effective data collection in a short duration of time.

Adequate sample size in qualitative research depends upon the judgment and experience of the researcher when evaluating the quality of data collection and deciding where to use judgment, the specific research method, purposeful sampling, and the research product intended (Merriam, 2009). A sample size might vary depending upon the complexity of a research. Morse (1994) recommended up to 50 participants for qualitative studies. We engaged two professionals in the entertainment industry to field-test study instrument by reviewing and providing feedback on clarity of each questionnaire interview questions. Scholars like Ekekwe (2013) used the expert validation strategy to ensure the reliability of study instruments by presenting interview questions to experts for their views of the instrument.

Letters and numbers were assigned to maintain anonymity and avoid questions that can compromise participant's confidentiality. Invitations were sent out for entertainment consumers located in NYC to take part in study through Mokeysurvey an online research website. To enhance trust, we explained the purpose and benefits of research to participants through the questionnaire. Furthermore, consent statement was attached with the questionnaire to indicate that participants understand the right to not participate or withdraw from study at anytime without consequences.

The participants were required to provide insights on the reason for purchases of replicate entertainment products. Data for this case study is saved on flash drives in the home office for 5 years to protect the rights of participants. Thereafter, data will be destroyed by shredding, crushing and burning appropriately. To ensure ethical protection of the participants, there is no access to raw data collection of the study to any induvidual not connected to research.

Research Method and Design

Method

Researchers use qualitative study to promotes the use of words rather than numbers and emphasizes the complexity characterizing human experience with the socio-cultural context in which humans act (Goussinsky, Reshef, Yanay-Ventura, & Yassour-Borochowitz, 2011). Castellan (2010) posited that a qualitative method is better suited for studies which involve interpreting the actions and interactions of participants. Deodhar, Saxena, Gupta, and Ruohonen (2012) found that researchers use qualitative methods to understand the construction of meaning through social interaction. Researchers can use the qualitative method to summarize themes, exploration, and words with the meanings in different contexts (Hunt, 2014). In addition, researchers use qualitative method to obtain flexible data collection, and place events in contexts (Verd & Andreu, 2011). In this study, we used the qualitative method as a means to explore and understand a social phenomenon (Trotter, 2012). Qualitative research method is appropriate for identifying perceptions of participants (Marshall & Rossman, 2010, 2011).

Qualitative researchers may use qualitative method, in identifying themes and constructs from language used by participants in response to research questions. Verner and Abdullah (2012) argued that individuals use qualitative research to make sense of the world. Qualitative research is abstractions that enable researchers give order to and categorize lived experiences (Verner & Abdullah, 2012). Doz (2011) posited that qualitative research uniquely suits for exploring institutional procedures. Verner and Abdullah noted that the qualitative approach entails a set of data collection and analysis techniques that researchers can use to generate a description, establish a theory, or build a theory.

Researchers could use the qualitative method to reveal a fuller picture in combination with a holistic approach to understanding a problem under study (Stake, 2010). In this study, qualitative method was useful for enhancing objectivity. Data collection was descriptive with participants' expression in words rather than numbers (Barrat, Choi, & Li, 2011). Using the qualitative research method, researchers work on assumptions, unlike the quantitative method, in which researchers use theories. According to Trotter (2012), a qualitative approach is idiographic in interpretation facilitating interpretion of data about specifics of a problem rather than through generalizations. However, qualitative method has inherent weaknesses that include subjection to researcher's biases and the possibility for influence in a particular direction, depending on how a researcher understands a problem (Murphy & Yielder, 2010). Nevertheless, we selected qualitative method because data collection consists of words rather than numbers (Castellan, 2010). In addition, the study sample is small and not a random selection which align with case studies.

In contrast, quantitative method is designed for researchers to generalize study findings and not require deep understanding of issues. Researchers identify quantitative research method with positivism, the belief that physical and social reality is independent of individual observers (Castellan, 2010). Murphy and Yielder (2010) suggested that quantitative methods do not facilitate in-depth understanding of a problem in a statistical analysis, creating difficulty to interpret the findings. The quantitative method is not appropriate in this study because the data collection is qualitative, not quantifiable data required in statistical analysis. Quantitative methods are suitable for research in which theories or hypotheses explain variables (Hunt, 2014). Quantitative method is most suitable for deductive approaches, where researchers use theories or hypotheses to justify variables, the purpose statement, and direction for research questions (Castellan, 2010).

Researchers employ mixed methods, to reduce over-reliance on statistical data for describingsocial occurrences and experiences that are mostly subjective in nature (Jogulu & Pansiri, 2011). Mixed method is useful for investigating attitudes and for objective measurement of subjectivity (Simons, 2013). Mixed method approach is useful to addressinherent flaws of quantitative and qualitative methods. The advantage of using a mixed method approach is the rich data collection which researchers use to produce a complete picture of the research while reducing biases inherent in qualitative and quantitative methods (Ostlund, Kidd, Wengstrom, & Dewar, 2011). However, the mixed method approach is not suitable for this study because of less time for the in-depth analysis. According to Halaweh (2012), researcher may use case studies to develop theories, but not to make inferences. Therefore, in this study we used the qualitative case-study method to ensure an in-depth analysis.

Research Design

A design is the blueprint by which individuals change some undesired reality into some desired reality (Chakrabarti, 2010). The most significant determinant of research design may be a research question. A study design is the description of how researchers conduct research including how researchers obtain data (Knight, 2010). Research design improves clarity of communication between authors and readers and serves as a roadmap to readers. In addition, it enhances the negotiation of methods section by authors and readers (Knight, 2010). The central research question for this study is a "What" question. Halaweh (2012), suggested that the use of why questions lead to a case study. However, what questions are rationales for exploratory studies (Yin, 2009). Hashim, Hashim, and Esa (2011) posited that qualitative case studies enable researchers to analyze particular cases in perfect settings. For this study, we chose multiple-case-study design to facilitate an understanding of real-life contemporary phenomenon in context (Ritvala & Salmi, 2011).

In case study, researchers may collect data from multiple levels, perspectives, and sources. However, researchers can generalize from one case if the case is useful for theory building and testing (Alex, Näslund, & Jasmand, 2012; Vissak, 2010). Researchers can use case designs deeply to investigate dynamic, experiential, complex processes and areas (Vissak, 2010). Case designs consist of contextually rich data used by researchers to study a focused phenomenon on real-life context, providing an in-depth understanding of the nature and complexity of a phenomenon (Alex et al., 2012). Case design is appropriate for achieving a goal of this study to investigate the phenomenon and gain an understanding of the meaning in contemporary context (Deodhar et al., 2012).

The focus was on behaviors which influence consumer purchases of replicate entertainment product in NYC. Unit of analysis defines the case study, unlike other designs, and can help in the identification of current literature to clarify the phenomenon under study (Barratt, Choi, & Li, 2011). For this study, the units of analysis are the views of entertainment product consumers in NYC. DeLuca, McEwen, and Keim (2010) used a qualitative case design to explore risk perceptions of undocumented immigrants and intent to cross into the United States repeatedly. Konstantakis, Palaigeorgiou, Siozos, and Tsoukalas (2010) used a qualitative design to study attitudes, behaviors, and corresponding reasoning of computer science students regarding software replicate in Greece. Datta (2011) used a case study design, to explore the complexity and social changes in innovation. Caple (2012) used a single-case design to gain insights and understanding of how people use trust to collaborate in a work environment. Zhengchuan, Qing, and Chenhhong (2013) used a case study method to examine six computer hackers in China.

In the present study, we used qualitative questionnaire approach, with closed ended interview questions following the central question. The analysis of participants' questionnaire responses led to identification of emergent themes (Apulu &Latham, 2011; Song, McCAlearney, Robbins, McCullough, & Smith, 2011). Halaweh (2012) noted that researchers generalize in single-case and multiple-case studies that apply to fundamental principles rather than to populations.

Other qualitative research designs considered for this study include phenomenological design, a strategy in which the researcher studies lived experiences of a phenomenon as described by participants (Cope, 2011; Freeman, & Lindsey, 2012; Spence, & Elliot, 2012). Phenomenology is the study of participants lived experience and understanding how this experience leads to the development of a worldview (Stierand & Dörfler, 2012). Essential to all phenomenological research is a rich description of lived experience, through the understanding of the concept of the life-world being the context of experiences (Stierand & Dörfler, 2012). Phenomenological research design is used by researchers to study the meaning of lived experiences of a group of people around a specific phenomenon (Hunt, 2014). Phenomenological design approach is not suitable for this study because the design is most appropriate for exploring lived experiences.

The grounded theory design is used by researchers to generate new theories, beyond descriptions of individual lived experiences (Marshall & Rossman, 2011). Grounded theory will not be applicable to this study because the aim is not to discover new theories (Mutshewa, 2010). Additional qualitative design considered not suitable for this study is ethnography design. Ethnography is the study of shared behavior, beliefs, and language of intact cultural group in the field in a prolonged time (Boden, Muller, &Nett, 2011). Ethnography design is not appropriate for this study because data collection was not from large population, and we had no plan to acts as observers while immersed in the day-to-day lives of the phenomenon. Researchers use narrative inquiry to compose, order life experiences, and have a core focus on the study of experience asexperience is lived (McMullen & Braithwaite, 2013). Narrative research design is not appropriate for this study because the design works best when subject is an individual or a small group (Hunt, 2014). Qualitative case study approach enabled us to describe how participants understand the issue of product replication.

Population and Sampling

The population for interview will be limited to the entertainment consumers who have purchase replicate products in NYC. According to Lohr (1999), convenient samples are sampling of a populations that is close to a researcher. Furthermore, trading activities, including buying and selling of entertainment DVD and CD products is particularly notable in NYC (N. Akaeze, 2016). We sent out 180 questionnaires to participants using convinient sample with the aim to get up to 50 responses in this study. The criteria for selection included that participants are above 18 years of age, and have purchased replicate entertainment products. Participants were not selected by chance but based on their knowledge and expertise which facilitated an understanding of research problem and question, as well as specific issues related to study (Bagheri, Yaghmaei, Ashktorab, & Zayari, 2012; Poulis, Poulis, & Plakoyiannaki, 2013).

Participants for this study are consumers, with a minimum of 2 years' experience in purchases of entertainment products in NYC. A smaller sample size enables researchers to analyze responses from participants in detail, facilitating the identification of themes (Benard, 2012). Draper and Swift (2010) suggested that a sample size of between 5 and 25 while Mores (1994) found a sample of up to 50 participants as suitable for data collection in qualitative research study. In a qualitative case study, a sample must be small to increase in-depth understanding of the problem under study by enabling the researcher to obtain detailed information (Marshall & Rossman, 2011; Trotter, 2012).

Ethical Research

In the process of collecting data, the researcher has to respect the participants and the locations where the research study is conducted (Hanson, Balmer, & Giardino, 2011). Yin (2011) suggested that researchers are responsible for solving ethical problems arising from a study. In addition, the researcher must take extra care to remove all risks to participants when conducting a study (Hanson et al., 2011). Informed consent is substantially significant as a concept of moral and lawful requirements that protect participants partaking in the study (Jeong et al., 2012). Any study with interfaces that includes a person requires informed consent (Lambert & Glacken, 2011).

The invitation to participate in this study included an explanation of the process involved for participating or withdrawing from the study. Each participant had the option to withdraw from study at any time he or she feel uncomfortable or lose the interest in participating. To exit the study, the individuals need not complete questionnaires or send in responses. We included the consent statement as part of the invitation to complete the questionnaire for this study. In the consent statement, we described the research study, presented brief background information, explained the process of participantion; and then discussed the study's purpose.

A common practice by researchers is to offer incentives for participation in a study. However, for this study there was no compensation for participation. In addition, the consent statement included that data from the study is kept in a safe place for 5 years to protect participants' rights. Data were saved on a computer hard drive, flash drives with password protection and locked in a home library for safekeeping. Then we secured computer hard drive and flash drives and locked in home safe for 5 years, with intention to destroy by shreding and burning appropriately. The names of the participants and the organizations for which participants work were anonymous. The consent statement on questionnaire was useful as agreement document for this study (see Appendix A).

Data Collection

Questionnaires are a most convenient and inexpensive way of gathering information from people and could be used to cover a large geographical area (Hunter, 2012). Questionnaires are good for reducing interviewer bias for quantitative or qualitative data collection. During data collection, with questionnaires there are no verbal or visual clues that could influence participant responses in any certain way. Questionnaires are applicable through post to a large number of people in different geographical areas. However, some participants may send questionnaire responses back while others may not respond. The objectives of a study are achieveable with a well defined and designed Questionnaire.

Qualitative questionnaires could be used to gather facts about people's beliefs, feelings, experiences in certain jobs, service offered, activities and so on. The questionnaire was designed in such a way that participants had freedom to express views in response to the question asked without any influence or clues from the interviewer. The questions consisted of closed ended questions which allowed respondents write either positive or negative responses.

Data gathered in this format is helpful if researchers seek to understand how people feel about certain issues. For example: experiences in using certain products, feelings about service offered by a restaurant and so on. This type of research method could be useful for companies who seek to understand the experiences and feelings of consumers who use certain products. Responses from participants could influence the company to change strategies in designing certain products to suit the needs of consumers quickly.

However, Qualitative questionnaires may not be helpful if the researchers are interested in quantifying and confirming hypotheses about certain occurrences (Harris & Brown, 2010). The good thing about Qualitative questionnaires is the flexibility and posibilty of wording in different ways to allow participants to give responses in their own words compared to a yes or no. The data collection was conducted by questionnaires with a target of certain sample of entertainment consumers in NYC.

Instruments

Instrumentation is critical in qualitative research (Chenail, 2011). In qualitative research, the researcher is considered as the primary data collection instrument and must identify any biases and assumptions before the start of a study (Yin, 2011). According to Hashim et al. (2011), the ultimate instrument in a case study is the inquirer managing the study. Yin (2011) further posited that early identification of assumptions and biases on the part of researchers is beneficial for the study.

Fink (2003) suggested that questionnaires are ideal strategies for qualitative research. Qualitative questionnaires are useful for depth and individual meaning to question of interest (Gratton & Jones, 2004). For this study, E-questionnaires offered an inexpensive, quick and convenient way to collect data. To get the best results from e-questionnaires, questionnaire recipients should be targeted carefully and the value of their potential contribution to the project should be emphasized. E-questionnaires should be convenient, quick, and easy to access, and set out in a way that encourages full and complete responses (Harris & Brown, 2010).

For data, we also used information from books and peer reviewed journal articles to triangulate questionnaire data collection from 50 consumers of entertainment products in NYC. A Likert psychometric scale commonly used for questionnaires was useful to measure respondents' attitudes from closed ended question or statement in this study. The participants were encouraged to specify their level of agreement to a statement. We advised that participants may withdraw from study at any time by simply ignoring the questionnaire or stop completion at any stage.

Data Collection Technique

Participation in this study was not limited to consumers of any branch of the entertainment production in NYC. We emphasized confidentiality through the questionnaire instruments to build participants' confidence and encourage participants to answer questions freely. The questionnaire was designed to answer central research questions relating to four major factors that influence consumer purchasing behaviors. Lee (2009) and Rani (2014) identified the factors as;

1) Cultural,
2) Social,
3) Personal,
4) Psychological.

Questionnaire which is a kind of written interview was useful for collection of data because we did not need to be present for participants to provide useful information. Questionnaires provided relatively cheap, quick and efficient way collect information from more than a few individual participants. The use of questionnaires saves researchers some time (Harris & Brown, 2010). For this study, the questionnaire consisted of closed ended questions structured to allow only answers which fit into categories that were decided in advance.

Standardized closed questions facilitated ordinal data which was then ranked using a rating scale to measure the strength of participants purchase behaviors. Data collection was quick because of the closed questions encouraging participant to provide answers by ticking a box. All participants are asked the same questions in the same order. This means a questionnaire can be replicated easily to check for reliability. Therefore, researchers can use the questionnaire to replicate research and ascertain consistency.

Data Organization Techniques

Journals used by researchers as a method of data collection are a valid method of accessing rich qualitative data (Hayman, Wilkes & Jackson, 2012). We used a log book in the collection of data, and for tracking all the questionnaire responses, data and documents (Yin, 2011). Journal was uyseful to record our understanding of the issue from participants' responses. Bases of findings for this study were the experiences of entertainment consumers while journaling was a method for data collection. Journal notes were maintained in chronological order by date and time (Hayman et al., 2012). Additional inclusion in the journal was all meanings that were ascribed to participants' responses (Hayman et al., 2012).

Letters and numbers were assigned to represent participants' responses in place of names or descriptions to ensure anonymity. All data were saved and stored in a safe for 5 years before the final destruction of the data. The completion of data collection, file organization, labeling, and categorizing data facilitated the identification of different categories and themes. Finally, we shared results of this study through journal publication and scholarly book. The qualitative questionnaire that was used to obtain data from participants is in Appendix A.

Data Analysis Technique

In this study, potential themes included culture, religions, racial groups, nationalities, wealth, education, occupation, reference groups, family, role and status, family, roles and status, lifestyle, economic situation, occupation, age, personality and self-concept. Other potential themes are dominance, aggressiveness, self-confidence, perception, motivation, learning, beliefs and attitudes. Additional potential themes are physiological needs, biological needs, social needs, selective attention, selective distortion and selective retention, product, price, place and promotion. There are diverse ways to analysis of the research data for this study.

According to Braun and Clarke (2006), thematic analysis is a qualitative analytic method for identifying, analyzing and reporting patterns (themes) within data. Thematic analysis is a standard data analysis method used by researchers in qualitative research to explore, determine, explain, and relate patterns in data (Petty, Thomson, & Stew, 2012). Thematic analysis involves identifying key themes and mapping out the relationships among ideas to create a thematic map (Petty et al., 2012). The analysis process is non linear because the researcher moves back and forth between data collection and analysis (Petty et al., 2012). Yin (2011) argued that the process of thematic analysis involves making sense of the texts and images from the data collection. For this study, we thematically conducted analysis to understand, represent, and interpret the questionnaire response data (Yin, 2011).

The analysis which Yin (2011) recommended involves the following steps; (a) transcribe interviews and records (b) review the transcribed notes to get the general meaning of the data, (c) code the data, by arranging data into manageable themes; and (d) explain the meaning of the casestudy. Nvivo trademarks software, a qualitative program for data analysis, expedites thematic coding and categorization of the data collection during the analysis stage (Bergin, 2011). Nvivo trademarks is useful to assist in the coding, referencing, counting, sorting, and displaying of data gathered from the participants

(Hanson et al., 2011). The Nvivo trademarks software program facilitates the search and identification for themes within a data collection.

Researchers may identify the issues through participants' repetitive responses, and analyze the data collection by noting repeated responses from participants as assigned to each theme. The data may then be noted and arranged in sequence of codes to generate interest in the whole data (Petty et al., 2012). Coded data could be collated using identifiable themes to match codes of the entire collection of data.

All identified themes and overall analysis of the themes are refined and final analysis performed by choosing compelling extracts from interview data. The analysis is connected to research question, related literature and conclusions reached with scholarly traits. The transcribed audio data and notes are entered into NVivo trademarks 11 software program (Reiter, Stewart, & Bruce, 2011; Rowley, 2012). The software facilitates identification and coding of emergent themes from the data. The thematic expressions from software analysis should support subsequent conclusions (Moustakas, 1994).

Van Kaam Method

For this study, we did not use modified van Kaam method which researchers normally combine with readings of transcribed material for assistance in organizing the output of interviews. Muto and Martin (2009) posited that Van Kaam produced a method for ensuring the effectiveness of qualitative research method. However, Moustakas (1994) provided a modified, seven-step, version of theVan Kaam method based upon individual textural-structural descriptions. The modified van Kaam method for research participant is as follows;

- listening and preliminary grouping;
- reduction and elimination to invariant constituents;
- clustering and thematizing of invariant constituents;
- final identification of invariant constituents and themes;
- individual textural description;
- individual structural description;
- and textural-structural description (Moustakas, 1994).

In alignment with the Whiteley (2012) the literature and conceptual idioms is addressed to which the emergent themes referred through critical and repeated examination of the thematic expressions with emergent coded themes, the reflexivity. Richards and Morse (2013) suggested the use of verbatim quotes from participants where necessary to support analysis of data. Then research should be located within current fields of study, and an argument presented for research into other fields by highlighting gaps and then full analyses and conclusions of research.

Questionnaire Responses Analysis Process and Procedure

For this study, we used a MonkeySurvey online questionnaire to collect half responses from participants, and the software automatically collated data. The study data was downloaded as a spreadsheet. Same questionnaire was later self-administered through social media online groups to complete the other half of responses. Responses were manually transferred from the questionnaires into a spreadsheet. Each question number was put as a column heading, and one row was used for each participant's answers. Then each possible answer was assigned a number or codes.

We went through each respondent's questionnaire in turn, adding in the codes. Data was then entered into a spreadsheet. See Appendix (B) below for what this look like. Upon completion of data entry from all the questionnaires into a spreadsheet, data was then cross checked for accuracy. There were no significant errors, and no need to re-check because all the data were present and correct.

We proceeded to calculate how many people selected each response by counting up manually to avoid input errors though it is easier to let the spreadsheet do the work, by adding a filter to each question within the spreadsheet. After calculating how many people selected each response, tables were set up to display the data. We looked at whether there was any variation in the way that different categories of participants responded. In particular, we looked at just the female responses, compared to just the male responses but found no significant variation. Once we completed analysis of all data, the findings and how to present what the data story is telling was discussed, along with the meaning in relation to central research questions.

Reliability and Validity

One of the primary difficulties in a qualitative study is substantiating reliability and certainty in the perception that an investigator presents to elucidate the phenomenon under investigation (Whiting & Sines, 2012). Drenth (2010) suggested thatproperly dealing with integrity and misconduct is particularly challenging. In seeking to establish trust, researchers may use a number of practical strategies that include letting participants guide the research, checking theoretical constructions against participants'definition of the phenomenon, and using participants' words in the theory (Whiting & Sines, 2012). Alqahtani (2011) posited that participants' validation enhances the credibility of a research study.

Four criteria that form the frame work for determining the rigor of research in a qualitative study are credibility, dependability, confirmability and transferability (Houghton, Casey, Shaw & Murphy, 2013). Credibility refers to the value and believability of research findings by conducting the research in a believable manner and being able to demonstrate credibility (Houghton, Casey, Shaw & Murphy, 2013). A key strategy to enhance credibility of qualitative research is maintenance of an audit trail (Ryan-Nicholls & Will, 2009). An audit trail is a collection of materials and notes used in the research process that documents a researcher's decisions and assumptions (Cope, 2014). For this study we maintained an audit trail to enhance credibility of the study.

Dependability compares to the concept of reliability in quantitative research and refers to data stability (Houghton, Casey, Shaw & Murphy, 2013). Confirmability has close links to dependability and refers to the neutrality and accuracy of the data. Transferability refers to whether or not researchers can transfer a particular finding to another similar context or situation, while still preserving the meanings and inferences from a completed study (Houghton et al., 2013). Transferability is the extent to which people realize a specified effect of a particular treatment in different research environment (Cambon, Minary, Ridde, & Alla, 2012). To enhance transferability, we adequately describe the original context of this research through thick description so that other researchers can make informed judgments (Houghton et al., 2013).

Reliability

Reliability is more concerning the external issues of a qualitative research (Sekaran, 2003). Golafshani (2003) posited that reliability of a research is related to generalizability of research result and suggested triangulation of the primary data to ensure reliability. Reliability is the ability to show that other researchers can replicate procedures used in a research study to achieve the same results (Dubois & Gilbert, 2010; Schiele & Krummaker, 2011). Issues that could potentially bias the reliability in this study included the wording of study questionnaire. Another issue is participant error where respondents could answer to appropriate their own image because product replication is an unconventional practice. Additional potential issue involves observer bias concerning the interpretation by the respondents.

We addressed all study issues through our choice of research method, design, numerated self-completion questionnaires where standardized questions and standardized customs of decoding and interpreting of data eliminated the risks. To ensure consistency, we secured the raw data for 5 years to address any inconsistency in the research and then discard the raw data afterwards. Furthermore, to establish reliability, we ensured that there was no deviation from research method selected for this study. Bias affects reliability, generalizability, and applicability of any research study (Murphy & Yielder, 2010). In addition, to maintain reliability in the study, we avoided using any leading questions, and gave participants the space to answer questions freely. In this study, primary data collection through closed ended research questionnaire was validated by triangulating with information from secondary data sources including books, business and academic journals and websites.

Validity

Validity in a qualitative study to a large extent is internal issues that could risk the trustworthiness (Sekaran, 2003). Internal validity in qualitative research is the establishment of accuracy or trustworthiness of the study from a standpoint of the researcher, participant, and reader (Yin, 2011). We used description of study findings, self-monitoring, or clarification of researcher's role, and an external auditor to review research to improve internal validity (Murphy & Yielder, 2010). Additionally, to enhance internal validity, we established chain of evidence in the data collection phase, by using notes and journals. In qualitative research, external validity means transferability or applicability of the findings to other individuals or sites of study with similar characteristics (Cambon et al, 2012; Petty, et al., 2012; Polit & Beck, 2010). Therefore, to establish the transferability (external validity) in this study, we defined the scope, the boundaries, and use convinient sampling. In this study, the scope was NYC in the United States.

Transition and Summary

Section 2 of this study consists of the purpose statement, role of the researcher, study participants, research methods, ethical procedures, and research design. In addition, Section 2 contains data collection instruments, the population, sampling methods and process, data organization, and data analysis techniques. Finally, Section 2 includes the research questions as well a description of reliability and validity of instruments. In Section 3, we presented the research findings, discussed the applications to professional practice, implications for social change, recommendation for action, recommendation for further study, and reflection on study esults.

SECTION 3

Application to Professional Practice and Implications for Change

Introduction

Section 3, consists of findings on behaviors of consumers which influences the purchases of replicate entertainment products. Section 3 also includes (a) an overview of the study, (b) presentation of findings, (c) application to professional practice, and (d) implication for social change. In addition, Section 3 includes recommendation for actions, recommendations for further study, reflections, and summary and study conclusion.

The purpose of this qualitative multiple case study is to explore consumer behaviors which influence complaisance towards purchasing replicate entertainment products. The problem was the demand for replicate products by consumers of entertainment products. The overraching research question for this study is: What consumer behaviors influence complaisance towards purchases of replicate entertainment products? The participants for this study consisted of 50 consumers of entertainment product who have purchased replicate entertainment products for not less than two years. Results of study relate to both the research question and conceptual framework.

Conceptual framework was TPB which researchers widely use to explain the practice of purchasing replicate products (Wiedmann et al., 2012). For this study, we conducted E-questionnaires to collect responses from consumers who purchased replicate entertainment products for not less than two years in NYC who answered to overarching research question. The sample was convenient with detailed information on participant's experiences after consenting to complete questionnaire. Participants responded to 31 closed-ended questions.

After organizing data, we analyzed the data for patterns and categorized themes. Names of the participants were not reqiured during data collection.

The study datawas classified into five categories: Cultural Influence (CFI), Social Influence (SFI), Personal Influence (PFI), Psychological Influence (PsFI) and Marketing mix Influence (MFI). Section 3 includes a review of findings from data collection for the multiple casestudy. The section also includes a summary of the data analysis, applications to professional practice, implications for social change, recommendations for further studies, and reflections on the study.

Presentation of the Findings

The TPB suggests that a person's behavior is determined by intention to perform a behavior; in turn, this intention is a function of attitude towards the behavior, subjective norm, and the perceived behavioral control (Ajzen, 1991). The concept of TPB details explanation on the factors of human behaviors which influences consumption of products. Participants responded to 31 close ended interview questions which we designed to answer the following research question: What consumer behaviors influence complaisance towards purchases of replicate entertainment products? The data analysis led to 13 themes based on highest number agreements by participants to question:

Cultural Factors
  2. Please indicate the frequency with which you have purchased replicate entertainment products in the past.
  3. Entertainment is an important part of my family life and activities.
  6. When I purchase products it is for my own preference
  7. I think social recognition is important to me.

Social Factors
  9. I became aware of replicate products through information from entertainment reference group or opinion leaders.
  10. My External sources of influence about replicate entertainment products includes Articles, reviews, advertising, or other activities of the entertainment company
  12. My Interpersonal sources of influence on purchasing replicate entertainment products include Opinions of friends, colleagues, relatives, or others.

Personal Factors
13. I always attempt to have a sense of accomplishment.
15. On average, how much money do you spend monthly on replicate entertainment products?
17. I'm self-conscious about the way I do things.

Psychological Factors
21. Purchasing replicate entertainment products generally benefits the consumer.

Marketing mix
29. In general, replicate entertainment products are more affordable.
30. In general, replicate entertainment products are more accessible.

Data collected for this study came from responses to closed ended interview questionnaires by 50 consumers of entertainment products. Data analysis of the questionnaire responses indicated 13 emerging themes, classed within five categories. The five main themes that emerged as well as the frequency and percentage of occurrence appear in Table 1. Three consumer behavior with the highest influences on purchases of replicate products were Personal influences, Cultural Influences, and Marketing mix Influences. The high percentages indicate that the influences are significant for influencing consumer decision towards purchases of replicate entertainment products.

Questionnaire Distribution and Response

The research topic was about entertainment consumer purchasing behavior, online questionnaire was appropriate for anonymity and protection of confidentiality. SurveyMonkey a websitewhich many researchers use for survey allowed participants to answer the questionnaire online. After creating the online questionnaire, we emailed the link to all participants. The respondents could easily click on the link which directly leads them to the questionnaire. To participate in this study, respondents were expected to have up to two years of experience in purchases of entertainment products. This condition was necessary because of the topic of the study which involved unconventional behaviors. The results were then recorded into an excel document by Spreadsheet. Link of the questionnaire was sent online between June 10th and June 25th, 2016.
A total of 180 questionnaires were distributed among the sample population. Because data collection instrument consisted of lengthy closed

ended questions, we expected a low turnover rate of up to 30% as suggested by Nulty (2008). Participation in the research was voluntary and respondents had a chance to stop and leave the questionnaire at any stage. Fifty three (29%) participants responded to the questionnaire, of which 5 (2.8%) were invalid and eliminated in line with the eligibility criteria and because of multiple responses. Only 48 (27%) eligible participant responses were collected. Therefore, 96% of our initial target research participants responses were valid data collection. For a copy of the questionnaire, see Appendix A.

From the data collection, 39(81%) of respondents said that their interpersonal sources of influence on purchasing replicate entertainment products includes Opinions of friends, colleagues, relatives, or others. 36(75%) of respondents either agreed or strongly agree that they were influenced by the opinions from friends or relative on replication. Furthermore, all 48(100%) of respondents said that they owned a replicate entertainment products; with 26(54%) of them rarely purchasing or sometimes purchasing the products and 22(46%) often or always purchasing replicate products. 46(95%) participants owned replicate products for their own preference; with 28(58%) strongly agreeing and 13(27%) also agreeing that they purchased products for their own preference while all (48) participants admitted that they are self-conscious about the way they do things. 43(`90%) respondents noted that in general, replicate entertainment products are more affordable while same number 43(`90%) declared that in general, replicate entertainment products are more accessible.

Components of the questionnaire

The questionnaire was designed with an introductory part which explains the purpose, content and consent to participate in the study. The questionnaire consists of 31 questions. Demographic questions including age, occupation, education and level, monthly income was part of the questionnaire. The questionnaire was distributed to individuals who purchased entertainment products and have at least one replicate product purchase experience. Respondents were asked why they purchased replicate entertainment products, what value they derive in purchasing products, and how much they spent.

In addition to these questions various statements were offered to respondents regarding their attitudes toward replicate products and asked to rate them on a Likert seven point scale from 1= strongly disagree and 7= strongly agree. Questionnaire form can be found in Appendix A. Secondary resource of data including raw data and published summaries, sources, such as, articles, books, journals, etc., facilitated initial insight into the research problem. In this research secondary data was useful for triangulation and were mainly collected from ProQuest Central and Google Scholar website (http://scholar.google.com) which provides so many articles and researches.

The coding process is to transform the raw data from the results of questionnaires into numerical data. In terms of the reliability of results, accuracy of raw data may be generated through computer program Statistical Package for Social Sciences (SPSS) which researchers widely adopt for analysis. Data for this study was presented in the form of tables, bar charts and following the figures was a detailed explanation of result.

Themes which emerged from analysis of the questinnaire data included: (a) Cultural Influence, (b) Social Influence, (c) Personal Influence, (d) Psychological Influence, and (e) Marketing mix Influence. Other subthemes also emerged under the five main themes.

The CFI included religions, racial groups, and nationalities influences. The SFI included reference groups, family, and, role and status. Under PFI, the subthemes that emerged were lifestyle, economic situation, occupation, age, personality and self-concept. PsFI included perception, motivation, learning, beliefs and attitudes influences. The MFI included product, price, place and promotion influences. Further analysis produced three major themes, Personal Influence, Cultural influence and Social influence which morphed from the five main themes.

Table 1

Frequency of Occurrence of Five Main Consumers Behavioral Influences for Purchasing Replicate Entertainmet Products

Themes	n	%
Cultural Influence	99	23
Personal Influence	152	35
Social Influence	98	23
Psychological Influence	25	5.8
Marketing mix Influence	57	13.2

Emergent Theme: Cultural Influences

Responses for Cultural **Influences** originated from closed ended questions number one to question seven in the questionnaire. Participant responses to Questions 2, 3, 6, and 7 explored the cultural behavioral factors which influences consumer purchases of replicate products. Cultures influence purchases and consumption of products as well as the derivable satisfaction (Ijewere, & Odia, 2012). The responses indicated that entertainment business managers should focus on Cultural Infuences fundamental implications for product development, pricing, distribution and promotion. Religions, racial groups, and nationalities influences are some the factors of Cultural influences which participants mentioned.

Of all respondents, forty 46 (96%) agreed that they purchase products for their own preference, 41(85%) said entertainment is an important part of their family life and activities, 34 (71%) think that social recognition is important to them, while 22(46%) respondents indicated high frequency of purchasing replicate entertainment products in the past (see Appendix B). lmost all participant 47 (98%) said they do not want to impress others when purchasing products. According to Ijewere and Odia (2012), culture impact on attitudes to things and issues like to foreign products, the types of dresses a married woman should wear, women drinking alcohol secretly or in public and time consciousness. Culture enables consumers to view products either as luxury and or a necessity.

Emergent Theme: Social Influence

Social influence is the affects of others on a consumer behavior (Haque, Khatibi & Rahman, 2009). This are either informational such as effect of the opinions of others regarding certain products and services on one's behavior or when an individual buys and tells the other individuals who also follow suit purchasing same products. The theme Social Influence originated from Questions 8, 11 and mainly Questions 9, 10, 12, with which we explored the participants' insight into social influence for purchases of replicate entertainment products.

Participants shared their thoughts about consumer behaviors that influence the decision to purchase replicate entertainment products. Lin, C. (2011) suggested that consumers establish their social status through purchases of products with symbolic implications. While all participants had a job, 39 (81%) said their Interpersonal sources of influence on purchasing replicate entertainment products includes Opinions of friends, colleagues, relatives, or others.

In addition, 35 (73%) of participants said their external sources of influence about replicate entertainment products included Articles, reviews, advertising, or other activities of the entertainment company while 32 (67%) of all participants agreed that their awareness of replicate products was through information from entertainment reference group or opinion leaders. On the influence of role and status in the society, participants were evenly split between agreement and disagreement with 26(54%) disagreed while 22(46%) agreed that their roles and status influence the purchases of replicate products.

Emergent Theme: Personal Influence

Personal Influence on consumer behaviors includes age, sex, occupation, life-style, personality (Purcarea & Rusanescu, 2011). The interaction between two or more people has a significant influence purchasing decisions. Through interpersonal interactions, individuals provide information to each other as well as attempt to influence purchasing outcome to derive benefit. The influence of one person on another is interpersonal influence which occurs within the context of a group (Purcarea & Rusanescu, 2011). The findings of this study indicated that PFI influences purchase decision of replicate entertainment products. Responses to Questions 13, 14, 15, 16, 17 and 18 which we asked participants to provide the relationship between Personal Factors, and the purchases of replicate provided insight on consumer behaviors and thinking. Respondents were a mixture of 32(67%) males and 16 (33%) females who are not less than 18 years of age.

All participants frequently purchased replicate entertainment products, however, only 3(~6%) spent significant amount on the products. The participants were split the question about how concerned they are about what other people think of them. While 21(44%) of participants said they are concerned, 27(56%) said they are less concerned about what other people think about them. Participants were also split on the question about their personality, with 27(56%) of participants agreeing that they are domineering, aggressive, or self-confidence but 21(44%) disagreed. All respondents 48(100%) said they are self-conscious about the way they do things, while 45(94%) of participants claimed that they always attempted to have a sense of accomplishment (see ----).

Emergent Theme: Psychological Influence

Psychological includes perception, motivation, learning, attitudes, and belief needs, perception, motives, attitudes, preferences, personalities, learning process and risks associated with purchasing (Roszkowska-Holysz, 2013). Under this theme, 27(56%) suggested that purchasing replicate entertainment products generally benefits the consumer while 28(58%) said that purchasing replicate entertainment products is not a better choice for consumers. 31(65%) of participants posited that the purchasing of replicate entertainment products is generally not convenient while 42(88%) admitted that they were not motivated to purchase replicate entertainment products because of a special needs. 26 (54%) rejected the idea that there is nothing wrong with purchasing replicate entertainment products while 36(75%) suggested that to buy or use counterfeits is not wise. Motives results from unsatisfied needs, which stimulate and guide the behaviors of consumers to satisfy the needs. Roszkowska-Holysz (2013) suggested that motives can trigger specific behaviors of consumers.

Emergent Theme: Marketing mix Influence

Marketing strategy is designed by managers to guide an enterprise use of resources in meeting the requirements of target customers and realizing marketing goals more efficiently than their competitors (C. Akaeze, 2016). The basic constituents of the Marketing mix include Products, Price, Place, and Promotion. Product is a tangible object or intangible service that is produced or manufactured and offered to consumers in the market. Price is normally an economic cost which is the amount a consumer pays for a product or service. Place or distribution channel includes physical stores as well as virtual outlets online or the location where a product or service can be purchased. Promotion is the communications that marketers use in the marketplace including advertising, public relations, personal selling and sales promotion (Gordon, 2012). In the current study, MFI derived from responses to Interview Questions 25, 26, 27, 28, 29 and 30, on participants insights about Marketing mix influence on consumer behavior towards purchases of replicate entertainment products. The purpose of the question was to explore aspects of the strategies which participants in a competitive environment used to gain and retain customers for success.

In general, 43(~90%) of participants suggested that, replicate entertainment products are more accessible and same number 43 (~90%) said that replicate entertainment products are more affordable. However, 25 (52%) of participants posited that overall, their positive experience with replicate entertainment products did not outweigh the negative experience. In addition, 28 (58%) suggested that they are not satisfied on their experience with purchasing replicate entertainment products while 26(54%) said that replicate entertainment products have no satisfying quality. Finally, 33(69%) of all participants admitted replicate entertainment products are not reliable. According to Gordon (2012), most of the past studies indicate that consumers have negative opinions towards marketing. By understanding the consumer sentiments entertainment managers may improve their products to better serve their consumers.

Table 2

Personal Influence

Themes	number of response	% of respondent agreement
lifestyle	39	9.0
economic situation	34	8.0
personality	39	9.0
and self-concept	22	5.0
Occupation	7	1.5
Age	11	2.5

Summary of Themes

Guba (1978) argued that when same concept reoccurs in a text, the concept is likely the theme and some of the most obvious themes in a collection of data regularly occurred and recurred. Further analysis of the data using participant's most affirmations to research questionnaire revealed three primary themes: (a) Personal Influence, (b) Cultural Influence and (c) Social Influence. Of the 13 emergent themes, participants mostly endorsed these three major themes as influence to the purchases of replicate products. Participants endorsed Personal Influence in the data 35% of the time, Cultural Influence 23% of the time and Social Influence 23% of the times.

This result supports Purcarea et al., (2011) who posited that Personal Influence of age, sex, occupation, life-style, personality influence consumer behaviors and differs from Muthiani and Wanjau (2012) who suggested price as the significant factor influencing product replication using of TPB. The interaction between two or more people has a significant influence purchasing decisions. Through interpersonal interactions, individuals provide information to each other as well as attempt to influence purchasing outcome to derive benefit. The influence of one person on another is interpersonal influence which also occurs within the context of a group

In addition, the findings regarding the Cultural Influence aligns with Ijewere and Odia (2012) findings which suggested that cultures influence purchases and consumption of products as well as the derivable satisfaction. Culture impact on attitudes to things and issues like to foreign products, the types of dresses a married woman should wear, women drinking alcohol secretly or in public and time consciousness. Culture enables consumers to view products either as luxury and or a necessity (Ijewere & Odia, 2012). Furthermore, the finding regarding Social Influence aligns with Lin, C. (2011) who suggested that consumers establish their social status through purchases of products with symbolic implications. According to Haque et al., (2009), Social influence is the affects others have on a consumer purchasing behavior. This can be informational such as effect of the opinions of others regarding certain products and services on one's behavior or when one person buys and tells the others who also follow suit purchasing same products.

The results for the themes originating from this study's data and analysis explain the entertainment consumer behaviors which influence the purchases of replicate products in NYC. The conceptual framework underlying this study was the TPB theory by Ajzen (1991). Consumer behavior is the study of when, why, how, and where people do or do not buy a product (Ajibola & Njogo, 2012). Consumer behavior and attitudes facilitates a better understanding and forecasting of products purchases and motives and frequency. Managers attempts to understand the buyer's decision making process, from understanding consumer behavior. Consumer behavior involves the characteristics of individuals such as demographics and behavioral which is useful to understanding people's wants.

Consumer behavior involves assessing the influences on consumers from family, friends, reference groups, and society in general (Ajibola & Njogo, 2012). Three significant entertainment consumer behaviors which influences purchasing replicate products identified in this study were Personal Influence, Cultural Influence and Social Influence. According to the participants, Personal Influences were most significant influence towards replicate entertainment products. In regards to consumer behavior, lifestyle, economic situation, occupation, age, personality and self-concept are important towards the purchases of replicate entertainment products. Consumers of entertainment products perceive Personal Influences as important influence on decision to purchase replicate entertainment products.

Applications to Professional Practice

A multiple case study is relevant to understanding the consumer behaviors which influence the purchases of replicate entertainment products. The purpose was to explore the strategies that owners of small auto dealership business use to survive in business within competitive environment beyond 5 years. The purpose of this qualitative multiple case study is to explore some consumer purchasing behaviors which influence complaisance towards replicate entertainment products. The findings section includes the evidence from participants, analysis of the data, and interpretation of the results.

The study offers insights to the behaviors which influence purchase decision of entertainment consumers, influences which managers need to improve managerial strategies. These behavioral influences are useful for entertainment managers to attract potential customers and retain existing consumers of entertainment products. Analysis of data revealed that some entertainment consumers complaisance to purchases of replicate products are influenced by Personal influences such as Personality, Lifestyle, Economic Situation, Occupation, Age. Consumers derive various benefits from purchases of replicate products including sense of accomplishment and social recognition through replicate entertainment products. They are attracted to accessibility, and affordability of replicate entertainment products.

The findings and recommendations might serve as the basis for entertainment business managers improve on the products, distribution, services, and customer satisfaction leading to development of the entertainment sector of the U.S. economy. The results could guide entertainment business managers who are struggling to survive because of product replication as to improve strategies and practices. The results offer insights into significant consumer behaviors which influences their purchases of replicate products. This study expands the body of literature for entertainment product replication. Some Entertainment corporation managers lack strategies to stop replication of entertainment products (N. Akaeze, 2016).

The findings from this study aligned with and impart the TPB. The TPB seeks to explain non-volitional behavior, goals and outcomes, which are not entirely under the control of the person. With willful behavior TPB argues that intention to perform a behavior is the best predictor of behavior. Intention is predicted by a person's attitude (e.g. whether a person sees a behavior as good, beneficial, pleasant, etc) and perceived social pressure to behave in a manner. The TPB includes a component of perceived control, with attitude and subjective norm, to predict behavioral intentions and may influence the intention-behavior link.

In line with the TPB concept, entertainment consumers purchase replicate products because they see it as beneficial, and because of perceived social pressures such as economic situation. The TPB is an excellent perspective for capturing consumer behavioral influences towards purchasing replicate entertainment products. Vida, Mateja, Kukar-Kinney and Penz (2012) indicated that deterrent strategies are inadequate to curb product piracy and suggested protecting intellectual property rights through legal strategies. Therefore, results indicate that doing business differently based on information from consumers may enable entertainment managers design new strategies to inhibit product replication.

Based on data analysis, entertainment managers need to make their products more affordable and accessible to all consumers. The data analysis showed that entertainment consumers do not see replicate products as better choice or more convenient. Majority entertainment consumers feel that replicate entertainment products are not reliable. They feel that the product quality are questionable and have less satisfying qualities. Entertainment product managers need to attract consumer using creative promotional techniques, improving logistics, giving offers and discounts to consumers and better quality product.

The findings were relevant to professional practice, as the study may provide information regarding how entertainment consumers make their purchase decision and what influences the decisions. The result could provide a practical guide for entertainment business managers to change business practices and improve business strategies that may curb the replication of products. The study's findings and recommendations add to the knowledge of business development by identifying the significant influences on entertainment consumer purchases for improved managerial strategies and decisions.

Implications for Social Change

According to Vogel (1998), the entertainment industry in the United States is responsible for $150 billion in expenditures and up to 120 billion hours of consumed time annually. However, successful entertainment business managers contribute a significant portion to the U.S. economy annually (N. Akaeze, 2016). The practice of entertainment product replication is growing at a rate of approximately 19% annually (N. Akaeze, 2016). This translates to revenue losses which in turn result in the termination of a significant number of entertainment businesses. Product replication affects the revenue of entertainment corporations through reduction of sales of original products. Product replication translates to loss of sales, income, and creative talent which may result in the collapse of the entertainment industry.

The knowledge acquired from understanding entertainment consumer behaviors which influences their purchases of replicate products may facilitate management success in inhibiting product replication. The information gained from this study may serve as a guide for entertainment corporation managers to curb product replication, increase sales of original products and opportunities to succeed. This research may guide entertainment managers in changing their plans, leadership, and management of their businesses. An improvement in the success rate of entertainment corporations may suffice for positive social change through increase in jobs, sales revenue, and creative talents. An improvement may also result in the reduction of unemployment rate and thereby creating successful, sustainable, and resilient businesses benefitting employees, their families, other businesses, communities, and the government.

Recommendations for Action

The intent of this study was to explore the consumer purchasing behaviors which influence complaisance towards replicate products with the aim to provide entertainment business managers a guide for strategies to curb replication. Small business owners are not adequately benefiting from value creation opportunities associated with their intangible assets (Laihonen & Lonnqvist, 2010). N. Akaeze (2016) argued that deterrent strategies have not been inadequate to curb product replication. Information from this study may serve as a resource for entertainment managers and a potential for entrepreneurial success. The participants provided valuable insights into their experience, knowledge, participation, procedures, and practices of product replication. Participant also revealed their perceptions concerning significant consumer behaviors leading to product replication practices which managers need to address for business success.

Consistent with previous studies, TPB was a strong predictor of consumer behavioral purchase intentions towards replicate entertainment products consumption. In addition, the role of personal influence on behavioral purchase of replicate entertainment products was demonstrated for our sample. The data obtained from this study provided entertainment business managers' insights for designing strategies towards curbing replication of product. Five recommended steps for action identified from this study should benefit current and future entertainment business managers to improve sales for growth and sustenance. Existing and potential entertainment business managers may concentrate products and marketing efforts toward addressing (a) Consumer Personal Influences, (b) Consumer cultural Influences, (c) Consumer Social Influences, (d) Marketing mix Influence, and (e) Psychological Influence.

Findings from the study indicated that products directed towards addressing entertainment consumers lifestyle, economic situation, occupation, age, personality and self-concept may lead to increase in purchasing of original products. Udo-Imeh (2015) showed that lifestyle significantly influence consumer buying behavior of college students in Nigeria. Income is a superior determinant of consumer purchasing behaviors (oszkowska-Holysz, 2013). The level of the obtained incomes affects the lifestyle and attitudes of a consumer. Therefore, manager's concerted efforts towards producing and marketing products designed to satisfy consumer needs influence their purchases and motivates them to purchase original products instead of replicates. We will disseminate the findings of the study through book publication, scholarly journals, and business journals. We will also present the findings at workshops and conferences, and offer written materials to Motion Pictures Association of America, and Director's guild of America.

Recommendations for Further Study

The purpose of the study was to explore the consumer purchasing behaviors which influence complaisance towards replicate products. The findings indicated that some consumer behaviors were significant for influencing the decision to purchase replicate entertainment products. Quantitative studies may expose a different account to the significance of these behaviors as influences to consumer purchases of replicate entertainment products. The study focused on consumers of entertainment products in NYC. Further study could provide useful information using entertainment consumers in different geographical location within the U.S. and extending to other countries, for better results and understanding of entertainment consumer purchasing behaviors for replicate products.

Further research can be carried out using more number of respondents. A valuable recommendation for further study is to explore how factors such as type of entertainment products influence the consumer purchases of replicate. Future researchers may carry out in-depth structured or semistructured interviews solely on entertainment consumers with experiences of replicate product consumption for in-depth analysis of their consumption behaviors. Analysis of their views may result to strategies aimed at converting to potential original consumers. The results of more research, when considering other variables, would provide entertainment managers with a broader analysis of consumer purchasing behaviors.

Reflections

Conducting the study has broadened our understanding of entertainment consumers in NYC. Some participants who are replicate product consumers were reluctant because of the perception that replication practice is an unconventional behaviors. Despite the challenges, data collection is valuable information for entertainment business leaders, and future researchers. We had no preconceived notions regarding the replicate entertainment consumption, participant's practices and research process. The data were through closed ended questionnaire to facilitate responses from experiences of study participants. All participants responded to each question and followed research proceedings in compliance with ethical research standards throughout the process. The study enhanced our understanding of consumer behaviors towards purchases of replicated entertainment products. Study results enhanced the understanding of entertainment businesses and stimulated interest to conduct further research on product replication.

Summary and Study Conclusions

Entertainment industry with the affiliate businesses contribute significantly to U.S. economy. In 2014 alone, movie industry sector of the entertainment industry released 697 movies products adding up to $10.4 billion the U.S. economy (Fox, 2015). Entertainment managers create jobs which include administration, management, marketing and entrepreneurship jobs (Kelman, 2015). The problem of entertainment business sector in the U.S. is product replication which results in losses of sale to entertainment corporations. According to De Vany and Walls (2007), replication practice is one of the most challenging problems faced by entertainment sector resulting in an estimated revenue loss of more than $3 billion annually to the U.S. studios. Deterrent strategies have not been inadequate to curb product replication (N. Akaeze, 2016). The purpose of this qualitative multiple case study was to explore the consumer purchasing behaviors which influence complaisance towards purchases of replicate products. The findings may serve as the basis for the development of business strategies for curbing product replication to improve sale leading to revenue growth in the U.S. The results could become a guide for entertainment business managers struggling on the successful strategies and practices to curb product replication and improve sales.

We collected data using closed ended questionnaire to gain insights of the consumer behaviors which influences their purchases of replicate entertainment products. Participants in this study were 50 participants who have purchased replicate product for more than two years and in NYC. The conceptual framework for this study was the TPB. Data collection had 13 emergent themes which we morphed under five categories (a) Consumer Personal Influences, (b) Consumer cultural Influences, (c) Consumer Social Influences, (d) Marketing mix Influence, and (e) Psychological Influence. Responses from participants indicated that Personal Influences was pivotal consumer behavioral influence to purchases of replicate products. Entertainment productions tailored toward addressing consumer personal influences may facilitate consumer purchases of original product and improve sales resulting in an increase in revenue for sustenance.

Participants submitted that, lifestyle significantly influence consumer buying behavior of entertainment consumers who purchase replicate products. Income was also a significant determinant of consumer purchasing behaviors for replicate products and pivotal towards competitive pricing and product differentiation strategies for managerial success. Participants provided insights into the purchasing behaviors of entertainment consumers which manager must pay attention to in order to succeed in business. The results could guide entertainment business managers who are struggling to curb replication of their products. Findings relate to TPB the conceptual framework of this study. The implication for positive social change includes the potential to reduce the facilitate strategies which managers may use to reduce product replication, improve sales leading to business sustenance, reduction of unemployment rate and crime.

The recommendation from this study consists of strategies that can benefit current and future entertainment business managers for sustainability and growth. In addition, this study's findings could provide managers with a practical guide to change entertainment business practices and improve the strategies that could promote sustainability and growth. We had the opportunity of collecting data from entertainment product consumers who purchased replicate products for more than 2 years in NYC. For a successful consumer oriented products, services and market, entertainment managers should work as psychologist to gain consumers. By keeping in mind information on consumer behaviors influencing the purchases managers may design favorable strategies to achieve consumer satisfaction. Study of consumer buying behaviour is gate way to success in production, distribution and marketing of entertainment. Conducting the study broadened our understanding of the purchasing behaviors of entertainment products consumers and strategies required by managers for success.

REFERENCES

Ahmad, T. (2010). Copyright infringement in cyberspace and network security: A threat to e-commerce. *IUP Journal of Cyber Law, 9*, 17-24. Retrieved from http://vufind.lib.bbk.ac.uk

Ajibola, O. D., & Njogo, B. O. (2012). The Effect of Consumer Behaviour and Attitudinal Tendencies Towards Purchase Decision (A case study of Unilever Nigeria PLC, Cadbury Nigeria PLC, United African Companies PLC. *Arabian Journal of Business and Management Review (Oman Chapter), 1*(12), 88-118. doi:10.12816/0002234

Ajzen, I. (1991). The Theory of Planned Behavior. Organizational Behavior and Human Decision Processes, 50(2), 179-211. doi:10.1016/0749-5978(91)90020-T

Ajzen, I. & Fishbein, M. (1975). Belief, Attitude, Intention, and Behavior: An Introduction to Theory and Research. Reading, MA: Addison-Wesley.

Akaeze, C. (2016). *Small Business Sustainability Strategies in Competitive Environments: Small Business and Competition.* Saarbrucken, Germany: Lambert Academic Publishing.

Akaeze, N. (2016). *Strategies Required by Managers to Inhibit Movie Piracy.* Saarbrucken, Germany: Lambert Academic Publishing.

Aleassa, H., Pearson, J., & McClurg, S. (2011). Investigating software piracy in Jordan: An extension of the theory of reasoned action.*Journal of Business Ethics, 98*, 663–676. doi:10.1007/s10551-010-0645-4

Alex, D., Näslund, D., & Jasmand, C. (2012). Logistics case study based research: Towards higher quality. *International Journal of Physical Distribution & Logistics Management, 42*, 275–295. doi:10.1108/09600031211225963

Alqahtani, A. (2011). The significance of English language learning in contemporary Kuwait: Some empirical insights for economists of knowledge and educational planning. *College Student Journal, 45*, 3–19. Retrieved from http://eric.ed.gov

Al-Rafee, S., & Dashti, A. (2012). A cross cultural comparison of the extended TPB: The case of digital piracy. *Journal of Global Information Technology Management, 15*, 5–24. Retrieved from http://www.uncg.edu

Alwehaibi, H. (2012). A proposed program to develop teaching for thinking in pre-service English language teachers. *English Language Teaching, 5*(7), 53–63. doi:10.5539/elt.v5n7p53

Andersen, B., & Frenz, M. (2010). Don't blame the P2P file-sharers: The impact of free music downloads on the purchase of music CDs in Canada. *Journal of Evolutionary Economics, 20,* 715–740. doi:10.1007/s00191-010-0173-5

Andrés, A., & Asongu, S. (2013). Fighting software piracy: Which governance tools matter. *Journal of Business Ethics, 118,* 667–682. doi:10.1007/s10551-013-1620-7

Apulu, I., & Latham, A. (2011). Drivers for information and communication technology adoption. A case study of Nigerian small and medium sized enterprises. *International Journal of Business Management, 6,* 260-268. doi:10.5539/ijbm.v6n5p51

Bagheri, H., Yaghmaei, F., Ashktorab, T., & Zayeri, F. (2012). Patient dignity and its related factors in the heart failure patients. *Nursing Ethics, 19,* 316-327. doi:10.1177/0969733011425970

Banutu-Gomez, M. (2013). The effects of piracy and counterfeiting on the international economy. *The Business Review, Cambridge, 21,* 33–43. Retrieved from http://jaabc.com

Barratt, M., Choi, T. Y., & Li, M. (2011). Qualitative case studies in operation management: Trends, research outcomes, and future research implications. *Journal of Operation Management, 29,* 329-342. doi:10.1016/j.jom.2010.06.002

Benard, H. R. (2012). *Social research methods: Qualitative and quantitative approaches.* Thousand Oaks, CA: Sage.

Bergin, M. (2011). NVivo 8 and consistency in data analysis: reflecting on the use of a qualitative data analysis program. *Nurse Researcher, 18*(3), 6-12. doi:10.7748/nr2011.04.18.3.6.c8457

Bernardi de Souza, F., & Sílvio, P. (2010). Theory of constraints contributions to outbound logistics. *Management Research Review, 33,* 683-700. doi:10.1108/01409171011055780

Bhaskaran, S., & Ramachandran, K. (2011). Managing technology selection and development risk in competitive environments. *Production and Operations Management, 20,* 541-555. doi:10.1111/j.1937-5956.2010.01165.x

Blankfield, S., & Stevenson, I. (2012). Towards a digital spine: The technological methods. *Publishing Research Quarterly, 28*(2), 79-92. doi:10.1007/s12109-012-9265-4

Blythe, J. (2008) The Official CIM Coursebook: Marketing Essential. UK: Elsevier

Boardman, M. (2011). Digital copyright protection and graduated response: A global perspective. *Loyola of Los Angeles International & Comparative Law Review, 33,* 223–254. Retrieved from http://digitalcommons.lmu.edu

Boden, A., Muller, C., & Nett, B. (2011). Conducting business ethnography in global software development project of small German enterprises. Information and Software Technology, *53,* 1012-1021. doi:10.1016/j.infsof.2011.01.009

Bonner, S., & O'Higgins, E. (2010). Music piracy: Ethical perspectives. *Management Decision, 48,* 1341–1354. doi:10.1108/00251741011082099

Booi, H., Chen, L., & Wilding, R. (2011). Managing production outsourcing risks in china's apparel industry: a case study of two apparel retailers. *Supply Chain Management, 16,* 428-445. doi:10.1108/13598541111171147

Braun, V., & Clarke, V. (2006). Using thematic analysis in psychology. *Qualitative Research in Psychology,* 3: 77-101. doi:10.1191/1478088706qp063oa

Bressi Nath, S., Alexander, L., & Solomon, P. (2012). Case managers' perspectives on the therapeutic alliance: a qualitative study. *Social Psychiatry and Psychiatric Epidemiology, 47,* 1815–1826. doi:10.1007/s00127-012-0483-z

Burke, D. (2010). The United States takes center stage in the international fight against online piracy & counterfeiting. *Houston Journal of International Law, 33,* 227–233. Retrieved from http://www.questia.com

Bustinza, O., Vendrell-Herrero, F., Parry, G., & Myrthianos, V. (2013). Music business models and piracy.*Industrial Management Data Systems, 113,* 4-22. doi:10.1108/02635571311289638

Cambon, L., Minary, L., Ridde, V., & Alla, F. (2012). Transferability of interventions in health education: a review. *BMC Public Health, 12,* 497. doi:10.1186/1471-2458-12-497

Cameron, R., & Molina, J. (2011). The acceptance of mixed methods in business and management research. *International Journal of Organizational Analysis, 19,* 256–271. doi:10.1108/19348831111149204

Caple, S. M. (2012). Trust re-defined a regional context. *Marketing Management Journal, 22,* 66-79. Retrieved from http://www.mmaglobal.org

Carlo, G., Padilla-Walker, L., & Day, R. (2011). A test of the economic strain model on adolescents' prosocial behaviors. *Journal of Research on Adolescence, 21,* 842–848. doi:10.1111/j.1532-7795.2011.00742.x

Carter, R., & Curry, D. (2013). Perceptions versus performance when managing extensions: new evidence about the role of fit between a parent brand and an extension. *Academy of Marketing Science. Journal, 41,* 253-269. doi:10.1007/s11747-011-0292-z

Castellan, C. (2010). Quantitative and qualitative research: A view for clarity. *International Journal of Education, 2*(2), 1-14. doi:10.5296/ije.v2i2.446

Castiglia, B. & Nunez, E. (2010). A Moral imperative-overcoming barriers to establishing an MBA infused ethics. *Journal of Legal, Ethical, & Regulatory issues, 13*(2), 33- 44. Retrieved from http://www.alliedacademies.org

Chakrabarti, A. (2010). A course for teaching design research methodology. *Artificial Intelligence for Engineering Design, Analysis and Manufacturing: AI EDAM, suppl. Design Pedagogy: Representations and Processes, 24,* 317-334. doi:10.1017/S0890060410000223

Chan, K., & Lai, M. (2011). Does ethical ideology affect software piracy attitude and behavior? An empirical investigation of computer users in China. *European Journal of Information Systems, 20,* 659–673. doi:10.1057/ejis.2011.31

Chenail, R. (2011). Interviewing the investigator: Strategies for addressing instrumentation and researcher bias concerns in qualitative research. *Qualitative Report, 16,* 255–262. Retrieved from http://www.nova.edu

Cheolho, Y. (2011). Theory of planned behavior and ethics theory in digital piracy: An integrated model. *Journal of Business Ethics, 100,* 405–417. doi:10.1007/s10551-010-0687-7

Cheung, N., & Cheung, Y. (2010). Strain, self-control, and gender differences in delinquency among Chinese adolescents: Extending general strain theory. *Sociological Perspectives, 53,* 321–345. doi:10.1525/sop.2010.53.3.321

Cheung, M., & Mcneil Boutté-queen, N. (2010). Assessing the relative importance of the child sexual abuse interview protocol items to assist child victims in abuse disclosure. *Journal of Family Violence, 25,* 11-22. doi:10.1007/s10896-009-9265-0

Chinaka, N. E. (2016). Factors that influence consumer purchasing behavior in Nigeria. The International Journal of Business & Management, 4(4), 157-161. Retrieved from http://www.theijbm.com

Cigdemoglu, C., Arslan, H. O., & Akay, H. (2011). A phenomenological study of instructors' experience on an open source learning management system. *Procedia-Social and Behavioral Science, 28,* 790-795. doi:10.1016/j.sbspro.2011.11.144

Cohen, A., & Avrahami, A. (2005). Soccer fans' motivation as a predictor of participation in soccer-related activities: An empirical examination in Israel. Social Behavior and Personality, 33(5), 419-434. doi.org/10.2224/sbp.2005.33.5.419

Coleman, L. J., Bahnan, N., Kelkar, M., & Curry, N. (2011). Walking the walk: How the theory of reasoned action explains adult and student intentions to go green. Journal of Applied Business Research, 27(3), 107-116. doi:10.19030/jabr.v27i3.4217

Collins, M., & Winrow, B. (2010). Porter's generic strategies as applied toward e-tailers post-leegin. *The Journal of Product and Brand Management, 19*, 306-311. doi:10.1108/10610421011059621

Connaughton, J., & Madsen, R. (2011). The economic impact of the film and video production and distribution industry on the charlotte regional economy. *Journal of Business & Economics Research, 9*(4), 15–26. Retrieved from http://www.cluteonline.com

Cope, Diane G. (2014). Methods and meanings: Credibility and trustworthiness of qualitative research. *Oncology Nursing Forum, 41*(1), 89-91. doi. org/10.1188/14.ONF.89-91

Cope, J. (2011). Entrepreneurial learning from failure: An interpretative phenomenological analysis. *Journal of Business Venturing, 26*, 604-623. doi:10.1016/j.jbusvent.2010.06.002

Cowart, T. (2011). The (final) anti-counterfeiting trade agreement (acta): Ruthless or toothless? *Southern Law Journal, 21*, 269-287. Retrieved from http://www.southernlawjournal.com

Crook, J. (2013). United States supports new treaty to facilitate visually impaired persons' access to books. *The American Journal of International Law, 107*, 933-934. Retrieved from http://www.asil.org

Cummings, A. (2010). From monopoly to intellectual property: Music piracy and the remaking of American copyright, 1909–1971. *Journal of American History, 97*, 659–681. Retrieved from http://www.academia. edu

Curry, L. A. (2009). Qualitative and mixed methods provide unique contributions to outcomes research. Circulation, 119, 1442-1452. doi:10.1161/CIRCULATIONAHA.107.742775

Dahlberg, L. (2011). Pirates, partisans, and politico-juridical space. *Law and Literature, 23*, 262–281, 295. doi:10.1525/lal.2011.23.2.262

Danaher, B., Dhanasobhon, S., Smith, M., & Telang, R. (2010). Converting pirates without cannibalizing purchasers: The impact of digital distribution on physical sales and Internet piracy. *Marketing Science, 29*, 1138–1151, 1166, and 1168. doi:10.1287/mksc.1100.0600

Daniel, L., Cassel, A., & Rodrigues, H. (2010). Service process analysis using process engineering and the theory of constraints thinking process. *Business Process Management Journal, 16*, 264-281. doi:10.1108/14637151011035598

Datta, P. B. (2011). Exploring the evolution of social innovation: A case study from India. *International Journal of Technology Management & Sustainable Development, 10*, 55-75. doi:10.1386/tmsd.10.1.55_1

Dave, D. R., & Patel, B. M. (2016). Impulsive Buying Behaviour in Organized Retail Stores With Specific Reference to FMCGS in Gujarat. *Prestige International Journal of Management and Research*, 8/9(2), 21-27. Retrieved from http://www.pimrindore.ac.in

De Castro, A., Gee, G., & Takeuchi, D. (2010). Examining alternative measures of social disadvantage among Asian Americans: The relevance of economic opportunity, subjective social status, and financial strain for health. *Journal of Immigrant and Minority Health*, *12*, 659–671. doi:10.1007/s10903-009-9258-3

Denscombe, M. (2010). The good research guide (4th ed.) New York, NY: McGraw Hill.

De Fina, A., & Perrino, S. (2011). Introduction: Interviews vs. "natural" contexts. *Language in Society*, *40*, 1–11. doi:10.1017/S0047404510000849

DeLuca, L., McEwen, M., & Keim, S. (2010). United states–Mexico border crossing: Experiences and risk. *Journal of Immigrant and Minority Health*, *12*, 113–123. doi:10.1007/s10903-008-9197-4

Deodhar, S. J., Saxena, K. B. C., Gupta, R. K., & Ruohonen, M. (2012). Strategies for software-based hybrid business models. *Journal of Strategic Information Systems*, *21*, 274-294. doi:10.10.1016/j.jsis.2012.06.001

De Vany, A.,S., & Walls, W. D. (2007). Estimating the effects of movie piracy on box-office revenue. *Review of Industrial Organization*, *30*(4), 291-301. doi:10.1007/s11151-007-9141-0

Dinh, T., & Helmarsson, H. (2013). Managing entry or expansion: Is Vietnam a feasible market for advanced food processing solution providers?. *Revista de Management Comparat International*, *14*, 567-584. Retrieved from http://www.rmci.ase.ro

Dolinski, J. (2012). Legal boundaries between internet piracy and a legal exchange of files through the internet. *Internal Security*, *4*, 165-180. Retrieved from http://www.isa.us.com

Downing, S. (2011). Retro gaming subculture and the social construction of a piracy ethic. *International Journal of Cyber Criminology*, *5*, 750–772. Retrieved from http://www.cybercrimejournal.com

Doz, Y. (2011). Qualitative research for international business. *Journal of International Business Studies*, *42*, 582–590. doi:10.1057/jibs.2011.18

Draper, A., & Swift, J. A. (2011). Qualitative research in nutrition and dietetics: data collection issues. *Journal of Human Nutrition and Dietetics*, *24*, 3-12. doi:10.1111/j.1365-277X.2010.01117.X

Drenth, P. (2010). Research integrity; protecting science. *European Review*, *18*, 417–426. doi:10.1017/S1062798710000104

Dubois, A., & Gilbert, M. (2010). From complexity to transparency: Managing the interplay between theory, method, and empirical phenomena IMM case studies. *Industrial Marketing Management, 39*, 129-136. doi:10.1016/j.indmarman.2009.08.003

Exum, M., Turner, M., & Hartman, J. (2012). Self-reported intentions to offend: All talk. *American Journal of Criminal Justice: AJCJ, 37*, 523-543. doi:10.1007/s12103-011-9148-9

Finks, A. (2003), The survey handbook, London: Sage publications.

Fox, M. A. (2015). The Economics of Drive-in Theatres: From Mainstream Entertainment to Nostalgia on the Margins. *Economics, Management and Financial Markets, 10*(3), 43-56. Retrieved from https://www.questia.com

Fradley, M. (2010). The contemporary Hollywood film industry/brand Hollywood: Selling entertainment in a global media age. *Film Quarterly, 63*(4), 80–81. doi:10.1525/FQ.2010.63.4.80

Freeman, S., & Lindsey, S. (2012). The effect of ethnic diversity on expatriate managers in their host country. *International Business Review, 21*, 253-268. doi:10.1016/j.ibusrev.2011.03.001

Glanz, K., Lewis, F. M., & Rimer, B. K. (Eds.). (1997). Health behavior and health education: Theory, research and practice (2nd ed.). San Francisco: Jossey-Bass

Goffe, L. (2011, February). Nollywood movies, counterfeiters beware! *New African, 503*, 70–71. Retrieved from http://newafricanmagazine.com

Golafshani, N. (2003), Understanding Reliability and Validity in Qualitative, *The Qualitative report*, 8(4), 597-607. Retrieved from http://www.nova.edu

Goldratt, E. M. (1994). *It's Not Luck*. Croton-on-Hudson, NY: North River Press, Inc.

Goode, S. (2010). Exploring the supply of pirate software for mobile devices: An analysis of software types and piracy groups. *Information Management & Computer Security, 18*, 204–225. doi:10.1108/09685221011079171

Gopal, R., & Gupta, A. (2010). Trading higher software piracy for higher profits: The case of phantom piracy. *Management Science, 56*, 1946–1962. doi:10.1287/mnsc.1100.1221

Gordon, R. (2012). Re-thinking and re-tooling the social marketing mix. *Australasian Marketing Journal, 20*(2), 122-126. doi.org/10.1016/j.ausmj.2011.10.005

Goussinsky, R., Reshef, A., Yanay-Ventura, G., & Yassour-Borochowitz, D. (2011). Teaching qualitative research for human services students: A three-phase model. *The Qualitative Report, 16,* 126-146. Retrieved from http://www.questia.com

Graebner, M. E., Martin, J. E., & Roundy, P. T. (2012). Qualitative data: Cooking without a recipe. Strategic Organization, 10, 276-284. doi:10.1177/147612701245281

Gratton, C. & Jones, I. (2004), Research methods for sports studies, London: Routeledge.

Grauerholz, L., & Bubriski-McKenzie, A. (2012). Teaching about consumption: The "not buying it" project. Teaching Sociology, 40(4), 332-348. doi:10.1177/0092055X12441713

Guba, E. G. 1978. Toward a methodology of naturalistic inquiry in educational evaluation. Monograph 8. Los Angeles: UCLA Center for the Study of Evaluation.

Gunter, W., Higgins, G., & Gealt, R. (2010). Pirating youth: Examining the correlates of digital music piracy among adolescents. *International Journal of Cyber Criminology, 4,* 657-671. Retrieved from http://www. cybercrimejournal.com

Gwyn, P., & Colin, J. (2010). Research with the doubly vulnerable: Population of individuals who abuse alcohol. *Journal of Psychosocial Nursing and Mental Health Services, 48*(2), 38–43. doi:10.3928/02793695-20100108-01

Halaweh, M. (2012). Integration of grounded theory and case study: An exemplary application from e-commerce security perception research. *Journal of Information Technology Theory and Application, 13,* 31–50. Retrieved from http://aisel.aisnet.org

Hanson, J. L, Balmer, D.F., & Giardino, A. P. (2011). Qualitative research methods for medical educators. *Academic Pediatrics, 11,* 375-386. doi:10.1016/j.acap.2011.05.001.

Haque, A., Rahman, S., & Khatibi, A. (2010). Factors Influencing Consumer Ethical Decision Making of Purchasing Pirated Software: Structural Equation Modeling on Malaysian Consumer. *Journal of International Business Ethics, 3*(1), 30-40. Retrieved from http://www.cibe.org.cn

Haque, A., Khatibi, A., & Rahman, S. (2009). Factors Influencing Buying Behavior of Piracy Products and its Impact to Malaysian Market. *International Review of Business Research Papers, 5*(2), 383-401. Retrieved from http://www.bizresearchpapers.com

Harris, Lois R. & Brown, Gavin T.L. (2010). Mixing interview and questionnaire methods: Practical problems in aligning data. Practical Assessment, Research & Evaluation, 15(1). Retrieved from: http://pareonline.net

Harrison, R., & Reilly, T. (2011). Mixed methods designs in marketing research. *Qualitative Market Research, 4,* 7-26. doi:10.1108/13522751111099300

Hashim, M., Hashim, Y., & Esa, A. (2011). Online learning interaction continuum (OLIC): A qualitative case study. *International Education Studies, 4*(2), 18–24. doi:10.5539/ies.v4n2p18

Hay, C., & Meldrum, R. (2010). Bullying victimization and adolescent self-harm: Testing hypotheses from general strain theory. *Journal of Youth and Adolescence, 39,* 446–459. doi:10.1007/s10964-009-9502-0

Hayman, B., Wilkes, L., & Jackson, D. (2012). Journaling: identification of challenges and reflection on strategies. *Nurse Researcher, 19*(3), 27-31. doi:10.7748/nr2012.04.19.3.27.c9056

Haynes, C. (2011). Piracy: The intellectual property wars from Gutenberg to gates. *Configurations, 19,* 143–146, 154. doi:10.1353/con.2011.0003

Hays, D., & Wood, C. (2011). Infusing qualitative traditions in counseling research designs. *Journal of Counseling &Development, 89,* 288–295. doi:10.1002/j.1556-6678.2011.tb00091.x

Hemming, J., Lordly, D., Glanville, N., Corby, L., & Thirsk, J. (2011). Developing an interview guide to evaluate practice-based evidence in nutrition: Use of the Delphi technique. *Canadian Journal of Dietetic Practice and Research, 72,* 186-90. doi:10.3148/72.4.2011.186

Higgins, G., & Marcum, C. (2012). Digital piracy: An integrated theoretical approach. *Choice, 49,* 1966. doi:10.5860/CHOICE.49–5953

Hinduja, S. (2012). General strain, self-control, and music piracy. *International Journal of Cyber Criminology, 6,* 951–967. Retrieved from http://www.cybercrimejournal.com

Ho, J., & Weinberg, C. (2011). Segmenting consumers of pirated movies. *Journal of Consumer Marketing, 28,* 252–260. doi:10.1108/07363761111143141

Hoffmann, J., & Spence, K. (2010). Who's to blame? Elaborating the role of attributions in general strain theory. *Western Criminology Review, 11*(3), 1–12. Retrieved from http://wcr.sonoma.edu

Hollander, S. (2011). Listen to the music: Lessons for publishers from record labels' digital debut decade. *Publishing Research Quarterly, 27,* 26–35. doi:10.1007/s12109-010-9192-1

Holma, A. M. (2012). Interpersonal interactions in business triads-case studies in corporate travel purchasing. *Journal of purchasing & Supply Management, 18*(2), 101-112. doi:10.1016/j.pursup.2012.04.002

Holt, T., & Copes, H. (2010). Transferring sub cultural knowledge on-line: Practices and beliefs of persistent digital pirates. *Deviant Behavior, 31,* 625–654. doi:10.1080/01639620903231548

Hougaard, J., & Tvede, M. (2010). Selling digital music: Business models for public goods. *Netnomics, 11,* 85–102. doi:10.1007/s11066-009-9047-0

Houghton, C., Casey, D., Shaw, D., & Murphy, K. (2013). Rigour in qualitative case-study research. *Nurse Researcher, 20*(4), 12-7. doi:10.7748/nr2013.03.20.4.12.e326

Hu, Y. (2011). How brand equity, marketing mix strategy and service quality affect customer loyalty: The case of retail chain stores in Taiwan. International Journal of Organizational Innovation (Online), 4(1), 59-73. Retrieved from http://www.ijoi-online.org

Hunt, L. (2014). In defense of qualitative research. Journal of Dental Hygiene (Online), 88(2), 64-5. Retrieved from http://jdh.adha.org

Hunter, Louise, MA, B.Sc, R.M. (2012). Challenging the reported disadvantages of e-questionnaires and addressing methodological issues of online data collection. Nurse Researcher (through 2013), 20(1), 11-20. doi:10.7748/nr2012.09.20.1.11.c9303

Hsiao-Chien, T., & Wang, T. (2012). Piracy and social norm of anti-piracy. *International Journal of Social Economics, 39*, 922-932. doi:10.1108/03068291211269361

Ifandoudas, P., & Gurd, B. (2010). Costing for decision-making in a theory of constraints environment. *Journal of Applied Management Accounting Research, 8*, 43-58. Retrieved from http://maaw.info/JAMAR.htm

Ijewere, A. A., & Odia, E. O. (2012). Cultural influences on product choice of the nigerian consumer. *Indian Journal of Economics and Business, 11*(1) Retrieved from http://www.serialsjournals.com

Jančić, D. (2010). The European political order and Internet piracy: Accidental or paradigmatic constitution-shaping? *European Constitutional Law Review, 6*, 430–461. doi:10.1017/S1574019610300058

Jaw, Y., Chen, C., & Chen, S. (2012). Managing innovation in the creative industries - a cultural production innovation perspective. *Innovation : Management, Policy & Practice, 14*, 256-275. doi:10.5172/impp.2012.14.2.256

Jens, H., & Mich, T. (2010). Selling digital music. *Business Models for Public Goods, 11*, 85–102. doi:10.1007/s11066-009-9047-0

Jeong, I., Kim, D., Kim, M., Kim, S., Jeong, D., & Shon, J. (2012). Exposure to and understanding of technical terms in informed consent forms for biomedical research. *Drug Information Journal, 46*, 19–26. doi:10.1177/0092861511427878

Jhr, W. (2009). Internet piracy and the European political and legal orders. *European Constitutional Law Review, 5*, 169–172. doi:10.1017/S1574019609001692

Jogulu, U., & Pansiri, J. (2011). Mixed methods: a research design for management doctoral dissertations. *Management Research Review, 34*, 687-701. doi:10.1108/01409171111136211

Kam, J., Cleveland, M., & Hecht, M. (2010). Applying general strain theory to examine perceived discrimination. *Prevention Science, 11*, 397–410. doi:10.1007/s11121-010-0180-7

Kelman, K. (2015). An entrepreneurial music industry education in secondary schooling: The emerging professional learning model. *MEIEA Journal, 15*(1), 147-173. Retrieved from http://www.meiea.org

Kigerl, A. (2013). Infringing Nations: Predicting Software Piracy Rates, BitTorrent Tracker Hosting, and P2P File Sharing Client Downloads Between Countries. *International Journal of Cyber Criminology, 7*, 62-80. Retrieved from http://www.cybercrimejournal.com

Klingenberg, B., & Watson, K. (2010). Intellectual property exchange between two partner companies - application of the theory of constraints thinking processes: [1]. *Journal of Business and Management, 16*, 125-138. Retrieved from http://www.chapman.edu

Klinger, B. (2010). Contraband cinema: Piracy, titanic, and central Asia. *Cinema Journal, 49*(2), 106–124. doi:10.1353/cj.0.0180

Knight, K. (2010). Study/experimental/research design: Much more than statistics. *Journal of Athletic Training, 45*, 98-100. doi:10.4085/1062-6050-45.1.98

Kohli, A., & Gupta, M. (2010). Improving operations strategy: Application of TOC principles in a small business. *Journal of Business & Economics Research, 8*(4), 37-45. Retrieved from http://cluteonline.com/journals

Koklic, M. (2011). Non-deceptive counterfeiting purchase behavior: Antecedents of attitudes and purchase intentions. *Journal of Applied Business Research, 27*(2), 127-137. Retrieved from http://www.cluteinstitute.com

Konstantakis, N., Palaigeorgiou, G., Siozos, P., & Tsoukalas, I. (2010). What do computer science students think about software piracy? *Behavior & Information Technology, 29*, 277–285. doi:10.1080/01449290902765076

Koster, A. (2012). Fighting internet piracy: The French experience with the hadopi law. *International Journal of Management & Information Systems, 16*, 327. Retrieved from http://cluteonline.com

Krstic, B., PhD., & Krstic, M., PhD. (2015). Rational Choice Theory and Random Behaviour. *Ekonomika, 61*(1), 1-13. doi:10.5937/ekonomika1501001K

Kurajdova, K., & Taborecka-Petrovicova, J. (2015). Literature review on factors influencing milk purchase behaviour. *International Review of Management and Marketing, 5*(1), 9-25. Retrieved from http://www.econjournals.com

Kureshi, S., & Sood, V. (2011). In-film placement trends: a comparative study of Bollywood and Hollywood. *Journal of Indian Business Research, 3*, 244-262. doi:10.1108/17554191111180591

Laird, B. (2011). Characterization of cancer-induced bone pain: an exploratory study. *Supportive Care in Cancer, 19,* 1393–1401. doi:10.1007/s00520-010-0961-3

Lambert, V., & Glacken, M. (2011). Engaging with children in research: Theoretical and practical implications of negotiating informed consent/assent. *Nursing Ethics, 18,* 781–801. doi:10.1177/0969733011401122

Larsson, S., Svensson, M., de Kaminski, M., Rönkkö, K., & Olsson, J. (2012). Law, norms, piracy and online anonymity. *Journal of Research in Interactive Marketing, 6,* 260-280. doi:10.1108/17505931211282391

Latuszynska, M., Furaiji, F., & Wawrzyniak, A. (2012). An empirical study of the factors influencing consumer behaviour in the electric appliances market. Contemporary Economics, 6(3), 76. doi:10.5709/ce.1897-9254.52

Lee, J. (2009). Understanding College Students' Purchase Behavior of Fashion Counterfeits: Fashion Consciousness, Public Self-Consciousness, Ethical Obligation, Ethical Judgment, and the Theory of Planned Behavior. (Electronic Thesis or Dissertation). Retrieved from https://etd.ohiolink.edu/

Letens, G., Farris, J., & Van-Aken, E. (2011). A multilevel framework for lean product development system design. *Engineering Management Journal, 23,* 69-85. doi: 10.1080/10429247.2011.11431887

Liao, C., Lin, H., & Liu, Y. (2010). Predicting the use of pirated software: A contingency model integrating perceived risk with the theory of planned behavior. *Journal of Business Ethics, 91,* 237–252. doi:10.1007/s10551-009-0081-5

Lin, C., Chi, Y., & Wang, C. (2012). Improvement strategies for logistics management - TOC approach. *Journal of Accounting, Finance & Management Strategy, 7*(2), 25-52. Retrieved from http://articles.lib.uts.edu.au

Lin, C. (2011). Personality, Value, Life Style and Postmodernism Consumer Behavior: A Comparison Among Three Generations. *International Journal of Organizational Innovation (Online), 3*(3), 203-230. Retrieved from http://www.ijoi-online.org

Lin, W. (2011, June 01). *General strain theory and juvenile delinquency: A cross-cultural study* (Unpublished doctoral dissertation). University of South Florida.

Lin, W. (2012). General strain theory in Taiwan: A latent growth. *Asian Journal of Criminology, 7,* 37-54. doi:10.1007/s11417-010-9101-8

Lin, W., Cochran, J., & Mieczkowski, T. (2011). Direct and vicarious violent victimization and juvenile delinquency: An application of general strain theory. *Sociological Inquiry, 81,* 195–222. doi:10.1111/j.1475-682X.2011.00368.x

Lin, W., & Mieczkowski, T. (2011). Subjective strains, conditioning factors, and juvenile delinquency: General strain theory in Taiwan. *Asian Journal of Criminology, 6,* 69–87. doi:10.1007/s11417-009-9082-7

Lind, F., Holmen, E., & Pedersen, A. C. (2012). Moving resources across permeable project boundaries in open network contexts. *Journal of Business Research, 65,* 177-185. doi:10.1016/j.jbusnes.2011.05.019

Lindgren, S., & Linde, J. (2012). The sub politics of online piracy: A Swedish case study. *Convergence, 18,* 143–164. doi:10.1177/1354856511433681

Lohr, S. L. (1999), *Sampling: Design And Analysis,* Pacific Grove, CA: Duxbury.

López-cuñat, J.,M., & Martínez-sánchez, F. (2015). Anti-piracy policy and quality differential in markets for information goods. European Journal of Law and Economics, 39(2), 375-401. doi:10.1007/s10657-013-9425-9

Lorde, T., Devonish, D., & Beckles, A. (2010). Real pirates of the Caribbean: Socio-psychological traits, the environment, personal ethics and the propensity for digital piracy in Barbados. *Journal of Eastern Caribbean Studies, 35,* 1–35. Retrieved from www.cavehill.uwi.edu

Luiz-Felipe, S., Reichhart, A., Hamacher, S., & Holweg, M. (2010). Managing product variety in emerging markets. *International Journal of Operations & Production Management, 30,* 205-224. doi:10.1108/01443571011018716

Lyons, M. (2010). Open access is almost here: Navigating through copyright, fair use, and the teach act., *41,* 65–66, 88. doi:10.3928/00220124-20100126-03

Mackenzie and Jurs (1993). Theory of reasoned action/theory of planned behavior, University of South Florida.

Maloy, J. S. (2008). A genealogy of rational choice: Rationalism, elitism, and democracy. Canadian Journal of Political Science, 41(3), 749-771. doi:10.1017/S0008423908080815

Marcum, C., Higgins, G., Wolfe, S., & Ricketts, M. (2011). Examining the intersection of self-control, peer association and neutralization in explaining digital piracy. *Western Criminology Review, 12*(3), 60–74. Retrieved from http://wcr.sonoma.edu

Marshall, C. & Rossman, G. (2011). *Designing qualitative research* (5th ed.). Thousand Oaks, CA: Sage.

Marsiglia, F., Kulis, S., Perez, H., & Bermudez-Parsai, M. (2011). Hopelessness, family stress, and depression among Mexican-heritage mothers in the southwest. *Health & Social Work, 36,* 7–18. doi:10.1093/hsw/36.1.7

Mcclennen, E. F. (2010). Rational choice and moral theory. Ethical Theory and Moral Practice, 13(5), 521-540. doi:10.1007/s10677-010-9253-8

McLennan, G., & Le, V. (2011). The effects of intellectual property rights violations on economic growth. *Modern Economy, 2,* 107–113. doi:10.4236/me.2011.22015

McManus, J. (2011). China's emerging software industry. *International Journal of Emerging Markets, 6,* 276–283. doi:10.1108/17468801111144021

McMullen, C., & Braithwaite, I. (2013). Narrative inquiry and the study of collaborative branding activity. *Electronic Journal of Business Research Methods, 11*(2), 92-104. Retrieved from http://www.ejbrm.com/main. html

Meissner, N. (2011). Forced pirates and the ethics of digital film. *Journal of Information, Communication and Ethics in Society, 9,* 195–205. doi:10.1108/1477996

Merriam, S. B. (2009). *Qualitative research: A guide to design and implementation.* San Francisco, CA: Jossey-Bass.

Meyer, T., & Leo, A. (2010). Graduated response and the emergence of a European surveillance society. *Journal of Policy, Regulation and Strategy for Telecommunications, Information and Media, 12,* 69–79. doi:10.1108/14636691011086053

Mirza, I. Z., & Islam, S. (2013). Patrolling the web for pirated content. Library Philosophy and Practice, 1-10. Retrieved from http://digitalcommons. unl.edu

Morris, R., & Higgins, G. (2010). Criminological theory in the digital age: The case of social learning theory and digital piracy. *Journal of Criminal Justice, 38,* 470–480. doi:10.1016/j.jcrimjus.2010.04.016

Moustakas, C. (1994). *Phenomenological research methods.* Thousand Oaks, CA: Sage.

Morse, J. M. (1994). Designing funded qualitative research. In Denizin, N. K. & Lincoln, Y. S., Handbook of qualitative research (2nd Ed). Thousand Oaks, CA: Sage.

Morrow, D. R. (2015). Wants and needs in mitigation policy. Climatic Change, 130(3), 335-345. doi:10.1007/s10584-014-1132-1

Murphy, F. J., & Yielder, J. (2010). Establishing rigor in qualitative radiography. *Radiography, 16,* 62-67. doi:10.1016/j.radi.2009.07.003

Mutshewa, A. (2010). The use of information by environmental planners: A qualitative study using grounded theory. *Information Processing and Management, 46,* 212-232. doi:10.1016/j.ipm.2009.09.006

Muthiani, M., & Wanjau, K. (2012). Piracy and social norm of anti-piracy. International Journal of Business and Social Science, 3(11), 87-96. Retrieved from http://ijbssnet.com

Naylor, G., & Susan, B. K. (2002). Exploring the differences in perceptions of satisfaction across lifestyle segments. Journal of Vacation Marketing, 8(4), 343-351. doi:10.1177/135676670200800405

Neenu, A., & Lobo, A. (2011, February 28). Piracy: Can movie, music industry really fight this menace? [hardware]. The Economic Times (Online) [New Delhi]. Retrieved from http://economictimes.indiatimes.com

Nga, J., & Lum, E. (2013). An investigation into unethical behavior intentions among undergraduate students: A Malaysian study. Journal of Academic Ethics, 11, 45-71. doi:10.1007/s10805-012-9176-1

Norazah, S., Ramayah, T., & Norbayah, S. (2011). Understanding consumer intention with respect to purchase and use of pirated software. Information Management & Computer Security, 19, 195–210. doi:10.1108/09685221111153564

Nulty, D. D. (2008). The adequacy of response rates to online and paper surveys: what can be done?. Assessment & Evaluation in Higher Education, 33(3), 301–314. Retrieved from https://www.uaf.edu

Oestreicher, K. (2011). Against the odds-The marketing dilemma of physical products in an increasingly virtual world. International Journal of Business and Social Science, 2(2), 39–53. Retrieved from http://www.ijbssnet.com

Oguer, F. (2011). The Hadopi Act vs. the global license as a psychological game. Review of European Studies, 3, 79–84. doi:10.5539/res.v3n1p79

Oliver, P. (2010). The DIY artist: issues of sustainability within local music scenes. Management Decision, 48, 1422-1432. doi:10.1108/00251741011082161

Orr, M. G., Thrush, R., & Plaut, D. C. (2013). The theory of reasoned action as parallel constraint satisfaction: Towards a dynamic computational model of health behavior. PLoS One, 8(5) doi:10.1371/journal.pone.0062490

Orser, B., Elliott, C., & Leck, J. (2011). Feminist attributes and entrepreneurial identity. Gender in Management, 26, 561-589. doi:10.1108/17542411111183884

Ostlund, U., Kidd, L, Wengstrom, Y., & Dewar, N. R. (2011). Combining qualitative and quantitative research with mixed method research designs. International Journal of Nursing Studies, 48, 369-383. doi:10.1016/j.inurstu.2010.10.005

oszkowska-Holysz, D. (2013). Determinants of consumer purchasing behaviour. Management, 17(1), 333-n/a. doi:10.2478/manment-2013-0023

Panas, E., & Ninni, E. (2011). Ethical decision making in electronic piracy: An explanatory model based on the diffusion of innovation theory and theory of planned behavior. *International Journal of Cyber Criminology, 5,* 836–859. Retrieved from http://www.cybercrimejournal.com/

Pang, L. (2010). Hong Kong cinema as a dialect cinema. *Cinema Journal, 49,* 140–143. doi:10.1353/cj.0.0197

Pan, W. (2010). Strategies for managing innovation in UK house building. *Engineering, Construction and Architectural Management, 17,* 78-88. doi:10.1108/09699981011011339

Patwardhan, M., Flora, P., & Gupta, A. (2010). Identification of secondary factors that influence consumer's buying behavior for soaps and chocolates. *IUP Journal of Marketing Management,* 9(1), 55-72. Retrieved from http://www.iupindia.in

Peerayuth, C., & Elkassabgi, (2011). The inverse u curve relationship between software piracy and technological outputs in developed nations. *Management Research Review, 34,* 968-979. doi:10.1108/01409171111158947

Petty, N. J., Thomson, O. P., & Stew, G. (2012). Ready for a paradigm shift? Part2: Introducing qualitative research methodologies and methods. *Manual Therapy, 17,* 378-384. doi:10.1016/j.math.2012.03.004

Phau, I., & Ng, J. (2010). Predictors of usage intentions of pirated software. *Journal of Business Ethics, 94,* 23–37. doi:10.1007/s10551-009-0247-1

Pi-Fang, H., & Miao-Hsueh, S. (2005). Using the theory of constraints to improve the identification and solution of managerial problems. *International Journal of Management, 22*(3), 415-425, 508. Retrieved from http://www.theijm.com

Piquero, N., Fox, K., Piquero, A., Capowich, G., & Mazerolle, P. (2010). Gender, general strain theory, negative emotions. *Journal of Youth and Adolescence, 39,* 380–392. doi:10.1007/s10964-009-9466-0

Polit, D. F., & Beck, C.T. (2010). Generalizability in quantitative and qualitative research: Myths and strategies. *International Journal of nursing Studies, 47,* 145-148. doi:10.1016/j.ijnurstu.2010.06.004

Poulis, K., Poulis, E., & Plakoyiannaki (2013). The role of context in case study selection: An international business perspective. *International Business Review, 22,* 304-314. doi:10.10.1016/jibusrev.2012.04.003

Pritchard, M., & Funk, D. (2010). The formation and effect of attitude importance in professional sport. *European Journal of Marketing, 44,* 1017-1036. doi:10.1108/03090561011047508

Purcarea, A. A., & Rusanescu, M. (2011). *Analysis of Differences in Purchasing Behavior of Individuals and Legal Entities and the Factors That Influence the Purchasing Behavior of Industrial Organizations.* Paper presented at the

International Conference on Management and Industrial Engineering, Bucharest. Bucharest: Niculescu Publishing House 151-157.

Qu, S., & Dumay, J. (2011). The qualitative research interview. *Qualitative Research in Accounting & Management, 8,* 238–264. doi:10.1108/11766091111162070

Rahman, S., Haque, A., & Rahman, M. (2011). Purchasing behavior for pirated products: A structural equation modeling approach on Bangladeshi consumers. *Journal of Management Research, 11,* 48–58. Retrieved from http://www.indianjournals.com

Rani, P. (2014). Factors influencing consumer behaviour. International Journal of Current Research and Academic Review, 2(9), 52-61. Retried from http://www.ijcrar.com

Reiter, S., Stewart, G., & Bruce, C. (2011). A strategy for delayed research method selection: Deciding between grounded theory and phenomenology. *Electronic Journal of Business Research Methods, 9*(1), 35-46. Retrieved from http://www.ejbrm.com/

Rhee, S., Cho, N., & Bae, H. (2010). Increasing the efficiency of business processes. *Information Systems Frontiers, 12,* 443-455. doi:10.1007/s10796-008-9145-9

Richards, L., & Morse, J. M. (2013). *Read me first for a user's guide to qualitative methods (3rd ed.).* Thousand Oaks, CA: Sage.

Ritvala, T., & Salmi, A. (2011). Network mobilize and target firms: The case of saving the Baltic Sea. *Industrial Marketing Management, 40,* 887-898. doi:1016/j.indmarman.2011.06.023

Robbins, W. (2011). Process improvement in the public sector: A case for the theory of constraints. *The Journal of Government Financial Management, 60*(2), 40-46. Retrieved from http://www.agacgfm.org/journal/

Robertson, K., McNeill, L., Green, J., & Roberts, C. (2012). Illegal downloading, ethical concern, and illegal behavior. *Journal of Business Ethics, 108,* 215-227. doi:10.1007/s10551-011-1079-3

Romeo, E. (2010). Quantitative research on critical thinking and predicting nursing students' nclex-rn performance. *Journal of Nursing Education, 49,* 378-386. doi:10.3928/01484834-20100331-05

Rosen, H. (2014, 09 02). Riaa mission statement. Retrieved from http://www.negativland.com

Rossman, G. (2010). Peer to peer and the music industry: The criminalization of sharing. *Contemporary Sociology, 39,* 691–692. doi:10.1177/0094306110386886i

Roszkowska-Holysz, D. (2013). Determinants of consumer purchasing behavior. *Management, 17*(1), 333-n/a. doi:10.2478/manment-2013-0023

Rowley, J. (2012). Conducting research interviews. *Management Research Review*, *35*, 260-271. doi:10.1108/01409171211210154

Ryan-Nicholls, K.D., & Will, C.I. (2009). Rigour in qualitative research: Mechanisms for control. *Nurse Researcher*, 16(3), 70-85. doi. org/10.7748/nr2009.04.16.3.70.c6947

Rybina, L. (2011). Music piracy in transitional post-soviet economies: Ethics, legislation, and expertise. *Eurasian Business Review*, *1*, 3–17. Retrieved from http://ideas.repec.org

Seedee, R. (2012). Moderating role of business strategies on the relationship between best business practices and firm performance. *International Journal of Business & Social Science*, *3*(24), 137-150. Retrieved from http://ijbssnet.com

Seidenberg, S. (2010). The record business blues. *ABA Journal*, *96*(6), 55–62. Retrieved from http://home.heinonline.org

Sekaran, U. (2003). Research Methods for Business: A Skill Building Approach (4th ed.). New York, NY. USA: John Wiley & Sons, Inc

Shah, R. B. (2014). Impact of marketing mix elements on customer loyalty: A study of fast food industry. *Prestige International Journal of Management and Research*, 6/7(2), 54-59. Retrieved from http://www.pimrindore. ac.in

Shamma, A. (2011). The use of mind mapping to develop writing skills in UAE schools. *Education, Business and Society: Contemporary Middle Eastern Issues*, *4*, 120–133. doi:10.1108/17537981111143855

Shaniel, D., & Bright, G. (2013). Advanced quality management system for product families in mass customization and reconfigurable manufacturing. *33*(2), 127-138. doi:10.1108/01445151311306636

Sharma, A. (2014). A study: Factors influencing purchase decision of consumer's for luxury products. *The International Journal of Business & Management*, 2(6), 161-172. Reteved from http://www.theijbm.com

Sheng, J., Shin, L., & Chou, C. (2010). Modeling the unethical intention of software piracy. *Quality & Quantity*, *44*, 191–198. doi:10.1007/ s11135-008-9188-5

Sherman, C., Tran, C., & Alves, Y. (2010). Elementary school classroom teacher delivered physical education: Costs, benefits and barriers. *Physical Educator*, *67*, 2–17. Retrieved from http://www.eric.ed.gov

Simburg, J., Astle, J., Bornhauser, J., Brushaber, S., & Fahlberg, S. (2012). International intellectual property. *The International Lawyer*, *46*, 215-230. Retrieved from http://ilra.law.smu.edu/Home.aspx

Simons, J. (2013). An introduction to q methodology. *Nurse Researcher*, *20*(3), 28-32. doi:10.7748/nr2013.01.20.3.28.c9494

Simon, R. (2012). The social construction of crime in the Atlantic world: Piracy as a case study. *The International Journal of Interdisciplinary Social Sciences, 6*, 75–88. Retrieved from http://iji.cgpublisher.com

Smallridge, J., & Roberts, J. (2013). Crime specific neutralizations: An empirical examination of four types of digital piracy. *International Journal of Cyber Criminology, 7*, 125-140. Retrieved from www.cybercrimejournal.com/

Song, P. H., McCAlearney, A. S., Robin, J., McCullough, J. S & Smith, B. T. (2011). Exploring the business case for ambulatory electronic health record system adoption/practitioner application. *Journal of Healthcare Management, 56*, 169-183. Retrieved from http://www.biomedsearch. com

Spence, B. D., & Elliot, E.A. (2012). African Micro entrepreneurship. The reality of everyday challenges. *Journal of Business Research, 65*, 1665-1673. doi:10.1016/j.jbusres.2012.02.007

Spink, J., & Fejes, Z. (2012). A review of the economic impact of counterfeiting and piracy methodologies and assessment of currently utilized estimates. *International Journal of Comparative and Applied Criminal Justice, 36*, 249-271. doi:10.1080/01924036.2012.726320

Spotts, H. (2010). We'd rather fight than switch: Music industry in a time of change. *Journal of the International Academy for Case Studies, 16*(5), 33-46. Retrieved from http://www.questia.com

Squires, J., Estabrooks, C., Gustavsson, P., & Wallin, L. (2011). Individual determinants of research utilization by nurses: A systematic review update. *Implementation Science, 6*, 1–20. doi:10.1186/1748-5908-6-1

Sreeroopa, S. (2013). *Qualitative methods: An example* (Unpublished raw data). Walden University, Minneapolis, MN.

Stake, R. E. (2010). *Qualitative research: Studying how things work.* New York, NY: The Guilford Press.

Stierand, M., & Dörfler, V. (2012). Reflecting on a phenomenological study of creativity and innovation in haute cuisine. *International Journal of Contemporary Hospitality Management, 24*, 946-957. doi:10.1108/09596111211247254

Street, C., & Ward, K. (2012). Improving validity and reliability in longitudinal case study time lines. *European Journal of Information Systems, suppl. Special Issue: Qualitative Research Methods, 21*, 160-175. doi:10.1057/ejis.2011.53

Suri, H. (2011). Purposeful sampling in qualitative research synthesis. *Qualitative Research Journal, 11*, 63–75. doi:10.3316/QRJ1102063

Tade, O., & Akinleye, B. (2012). 'We are promoters not pirates': A qualitative analysis of artistes and pirates on music piracy in Nigeria. *International*

Journal of Cyber Criminology, 6, 1014–1029. Retrieved from http://www.cybercrimejournal.com

Talmy, S. (2010). Qualitative interviews in applied linguistics: From research instrument to social practice. *Annual Review of Applied Linguistics, 30,* 128–148. doi:10.1017/S0267190510000085

Tao, S. (2013). Personality, motivation, and behavioral intention in the experiential consumption of artworks. Social Behavior and Personality, 41(9), 1533-1546.

Taylor, S. (2012). Evaluating digital piracy intentions on behaviors. *The Journal of Services Marketing, 26,* 472-483. doi:10.1108/08876041211266404

Tolbert, S., Moore, G., & Wood, C. (2010). Not-for-profit organizations and for-profit businesses: Perceptions and reality. *Journal of Business & Economics Research, 8*(5), 141-153. Retrieved from http://journals.cluteonline.com

Tokuyama, S., & Greenwell, T. C. (2011). Examining similarities and differences in consumer motivation for playing and watching soccer. *Sport Marketing Quarterly,* 20(3), 148-156. Retrieved from http://fitpublishing.com

Trotter II, R. T. (2012). Qualitative research sample design and sample size: Resolving and unresolved issues and inferential imperatives. *Preventive Medicine, 55,* 398-400. doi:10.1016/j.ypmed.2012.07.003.

Turri. Smith, & Kemp, (2013). Developing affective brand commitment through social media. *Journal of Electronic Commerce Research, 14,* 201-214. Retrieved from http://www.csulb.edu/journals/jecr/

Turner, W. (2010). Qualitative interview design: A practical guide for novice investigators. *The Qualitative Report, 15,* 754–760. Retrieved from http://mediaclass.co.uk/

Uchenna, U. J. (2015). *The Impact of Consumer Behaviour and Factors Affecting on Purchasing Decisions.* Paper presented at the Global Conference on Business & Finance Proceedings: Institute for Business & Finance Research, Hilo, 10(1) 204-212. Udo-Imeh, P. (2015). Influence of Lifestyle on the Buying Behaviours of Undergraduate Students in Universities in Cross River State, Nigeria. *Australian Journal of Business and Management Research, 5*(1), 1-10. Retrieved from http://www.ajbmr.com/current-issue/61

Valipour, H., Birjandi, H., & Honarbakhsh, S. (2012). The effects of cost leadership strategy and product differentiation strategy on the performance of firms. *Journal of Asian Business Strategy, 2,* 14-23. Retrieved from http://www.aessweb.com

Verner, J. M., & Abdullah, L. M. (2012). Exploratory case study research: Outsourced project failure. *Information and Software Technology, 54*, 866-886. doi:10.1016/j.infsof.2011.11.001

Vichak, P., & Johri, L. (2011). Impact of business strategies of automobile manufacturers in Thailand. *International Journal of Emerging Markets, 6*, 17-37. doi:10.1108/17468801111104359

Vida, I., Mateja, K., Kukar-Kinney, M., & Penz, E. (2012). Predicting consumer digital piracy behavior. *Research in Interactive Marketing, 6*, 298-313. doi:10.1108/17505931211282418

Vissak, T. (2010). Recommendations for using the case study method in international business research. *The Qualitative Report, 15*, 370-388. Retrieved from http://www.nova.edu/ssss/QR/index.html

Vogel, Harold L. 1998. *Entertainment Industry Economics: A Guide for Financial Analysis.* 4th ed. New York: Cambridge University Press.

Walls, W., & Harvey, P. (2010). DVD movie piracy in Hong Kong: Autopsy of a brick-and-mortar market. *International Journal of Management, 27*, 31–36, 200. Retrieved from http://ideas.repec.org/p/clg/wpaper/2009-19.html

Wan, W. W., N., Luk, C., Yau, O. H., M., Tse, A. C., . . . M. (2009). Do traditional Chinese cultural values nourish a market for pirated CDs? Journal of Business Ethics, 88, 185-196. doi:10.1007/s10551-008-9821-1

Watson, W., & Dow, K. (2010). Auditing operational compliance: The case of employee long distance piracy. *Issues in Accounting Education, 25*, 513–526. doi:10.2308/iace.2010.25.3.513

Waziri, K. (2011). Intellectual property piracy and counterfeiting in Nigeria: The impending economic and social conundrum. *Journal of Politics and Law, 4*, 196-202. doi:10.5539/jpl.v4n2p196

Welford, C., Murphy, K., & Casey, D. (2012). Demystifying nursing research terminology: Part 2. *Nurse Researcher, 19*(2), 29–35. Retrieved from http://europepmc.org/

Wells, W. D., Burnett, J., & Moriarty, S. (2005). Advertising: Principles and practice (7th ed.). Englewood Cliff, NJ: Prentice Hall.

Whiteley, A. (2012). Supervisory conversations on rigour and interpretive research. *Qualitative Research Journal, 12*, 251-271. doi:10.1108/14439881211248383

Whiting, M., & Sines, D. (2012). Mind maps: Establishing 'trustworthiness' in qualitative research. *Nurse Researcher, 20*, 21–27. Retrieved from http://www.ncbi.nlm.nih.gov/pubmed/23061270

Wiedmann, K., Hennigs, N., & Klarmann, C. (2012). Luxury consumption in the trade-off between genuine and counterfeit goods: What are the consumers' underlying motives and value-based drivers? Journal of Brand Management, 19(7), 544-566. doi:10.1057/bm.2012.10

Williams, P., Nicholas, D., & Rowlands, I. (2010). The attitudes and behaviors of illegal downloader's. *ASLIB Proceedings, 62,* 283–301. doi:10.1108/00012531011046916

Wing, M. (2012). The digital copyright time bomb in the BRIC economies, some ideas from the UK for the Indian market. *International Journal of Law and Management, 54,* 302-310. doi:10.1108/17542431211245332

Wohlers, A. (2012). Digital piracy: an integrated theoretical approach. *49,* 1966. doi:10.5860/CHOICE.49–5953

Wolfson, A. (2001). The costs and benefits of cost-benefit analysis. *Public Interest,* (145), 93-99. Retrieved from http://www.nationalaffairs.com

Woolley, D. (2010). The cynical pirate: How cynicism effects music piracy. *Academy of Information and Management Sciences Journal, 13,* 31–43. Retrieved from http://alliedacademies.org

Wu, H., Chou, C., Hao-Ren, K., & Mei-Hung, W. (2010). College students' misunderstandings about copyright laws for digital library resources. *The Electronic Library, 28,* 197-209. doi:10.1108/02640471011033576

Yin, R. (2009). *Case study research: Design and methods* (4th ed.). Thousand Oaks, CA: Sage.

Yin, R. (2010). *Case study research: Design and methods* (6th ed.). Thousand Oaks, CA: Sage.

Yin, R. K. (2011). *Qualitative research from start to finish.* New York, NY: The Guilford press.

Yoon, C. (2010). Theory of planned behavior and ethics. *Journal of Business Ethics, 100,* 405–417. doi:10.1007/s10551-010-0687-7

Yu-Chin, L., & Hsu, Y. (2013). Predicting adolescent deviant behaviors through data mining techniques. *Journal of Educational Technology & Society,* 16(1), 295-308 Retrieved from http://www.ifets.info/issues.php?id=58

Yu, S. (2010). Digital piracy and stealing: A comparison on criminal propensity. *International Journal of Criminal Justice Sciences, 5,* 239–250. Retrieved from http://www.ijcjs.co.nr/

Zhang, J., Hong, L. J., & Zhang, R. Q. (2012). Fighting strategies in a market with counterfeits. Annals of Operations Research, 192(1), 49-66. doi:10.1007/s10479-010-0768-0

Zhengchuan, X., Qing, H., & Chenhhong, Z. (2013). Why computer talents become computer hackers. *Communications of the ACM, 56*(4), 64–74. doi:10.1145/2436256.2436272

APPENDIX A

Questionnaire on Consumer Behaviors for replicate entertainment products

- **Please note that you may take:**
- Replication as piracy, bootlegging, counterfeiting, or faking etc.
- Entertainment products may be CD's, DVD's, downloading, peer 2 peer sharing etc.
- Send completed questionnaires to **krisakinfo@yahoo.com**

Replace 0 with x in preferred boxes to indicate your choices

NOTE: Completion of this questionnaire implies your consent to use your data. Your response will help me to understand Entertainment consumers purchase behavior of pirated products. You must be 18 or older to participate in this Questionnaire. Replicate entertainment products are unauthorized, reproduced copies entertainment products (e.g., movies, music, software, etc.).

Cultural Factors

Culture, subculture, and social class

1. **Academic qualifications**

   A. **Elementary education (0)**
   B. **Junior High School (0)**
   C. **Senior High School (0)**
   D. **Associate degree (0)**
   E. **College Graduate (0)**
   F. **Post graduate degree (0)**

2. **Please indicate the frequency with which you have purchased replicate entertainment products in the past.**

    A- **1 Never (0)**
    B- **2 Rarely (0)**
    C- **3 Sometimes (0)**
    D- **4 Often (0)**
    E- **5 Almost always (0)**
    F. **6 Always (0)**

3. **Entertainment is an important part of my family life and activities.**

    A- **Strongly agree (0)**
    B- **Agree (0)**
    C- **Slightly agree (0)**
    D- **Slightly disagree (0)**
    E- **Disagree (0)**
    F- **Strongly disagree (0)**

4. **I plan to purchase replicate entertainment products within the next month.**

    A- **1 Strongly agree (0)**
    A- **2 Agree (0)**
    B- **3 Slightly agree (0)**
    C- **4 Slightly disagree (0)**
    D- **5 Disagree (0)**
    E- **6 Strongly disagree (0)**

5. **When I purchase products I want to impress others**

    A- **Strongly agree (0)**
    B- **Agree (0)**
    C- **Slightly agree (0)**
    D- **Slightly disagree (0)**
    E- **Disagree (0)**
    F- **Strongly disagree (0)**

6.   **When I purchase products it is for my own preference**

A-   **Strongly agree (0)**
B-   **Agree (0)**
C-   **Slightly agree (0)**
D-   **Slightly disagree (0)**
E-   **Disagree (0)**
F-   **Strongly disagree (0)**

7.   **I think social recognition is important to me.**

A-   **1 Strongly agree (0)**
B-   **2 Agree (0)**
C-   **3 Slightly agree (0)**
D-   **4 Slightly disagree (0)**
E-   **5 Disagree (0)**
F-   **6 Strongly disagree (0)**

Social Factors

Roles and Status, Family, Reference Groups

8.   **Work status**

A.   **Do not work (0)**
B.   **Work less than 10 hours per week (0)**
C.   **Work 11-20 hour per week (0)**
D.   **Work 21-30 hour per week (0)**
E.   **Work 31-40 hours per week (0)**
F.   **Work 41-45 hour per week (0)**

9.   **I became aware of replicate products through information from entertainment reference group or opinion leaders.**

A-   **1 Strongly agree (0)**
B-   **2 Agree (0)**
C-   **3 Slightly agree (0)**
D-   **4 Slightly disagree (0)**
E-   **5 Disagree (0)**
F-   **6 Strongly disagree (0)**

10. **My External sources of influence about replicate entertainment products includes Articles, reviews, advertising, or other activities of the entertainment company**

   A- 1 Strongly agree (0)
   B- 2 Agree (0)
   C- 3 Slightly agree (0)
   D- 4 Slightly disagree (0)
   E- 5 Disagree (0)
   F- 6 Strongly disagree (0)

11. **My entertainment products buying decisions are mainly influenced by my role and status in the society.**

   A- 1 Strongly agree (0)
   B- 2 Agree (0)
   C- 3 Slightly agree (0)
   D- 4 Slightly disagree (0)
   E- 5 Disagree (0)
   F- 6 Strongly disagree (0)

12. **My Interpersonal sources of influence on purchasing replicate entertainment products includes Opinions of friends, colleagues, relatives, or others.**

   A- Strongly agree (0)
   B- Agree (0)
   C- Slightly agree (0)
   D- Slightly disagree (0)
   E- Disagree (0)
   F- Strongly disagree (0)

Personal Factors

Personality, Lifestyle, Economic Situation, Occupation, Age

13. I always attempt to have a sense of accomplishment.

    A- 1 Strongly agree (0)
    B- 2 Agree (0)
    C- 3 Slightly agree (0)
    D- 4 Slightly disagree (0)
    E- 5 Disagree (0)
    F- 6 Strongly disagree (0)

14. Age Group*

    A. 18-24 (0)
    B. 25-30 (0)
    C. 31-40 (0)
    D. 41-50 (0)
    E. 51-65 (0)
    F. Over 65 (0)

15. On average, how much money do you spend monthly on replicate entertainment products?

    A. Under $100 (0)
    B. $100 to $199 (0)
    C. $200 to $299 (0)
    D. $300 to $399 (0)
    E. $400 to $499 (0)
    F. $500 or more (0)

- **Gender**

    A. **Male (0)**
    B. **Female (0)**

16. **I'm concerned about what other people think of me.**

    A- **1 Strongly agree (0)**
    B- **2 Agree (0)**
    C- **3 Slightly agree (0)**
    D- **4 Slightly disagree (0)**
    E- **5 Disagree (0)**
    F- **6 Strongly disagree (0)**

17. **I'm self-conscious about the way I do things.**

    A- **1 Strongly agree (0)**
    B- **2 Agree (0)**
    C- **3 Slightly agree (0)**
    D- **4 Slightly disagree (0)**
    E- **5 Disagree (0)**
    F- **6 Strongly disagree (0)**

18. **I have a great personality because I am domineering, aggressive, or self-confidence**

    A- **1 Strongly agree (0)**
    B- **2 Agree (0)**
    C- **3 Slightly agree (0)**
    D- **4 Slightly disagree (0)**
    E- **5 Disagree (0)**
    F- **6 Strongly disagree (0)**

Psychological Factors

Perception, motivation, learning, beliefs and attitudes

19. There's nothing wrong with purchasing replicate entertainment products.

    A- 1 Strongly agree (0)
    B- 2 Agree (0)
    C- 3 Slightly agree (0)
    D- 4 Slightly disagree (0)
    E- 5 Disagree (0)
    F- 6 Strongly disagree (0)

20. Generally speaking, purchasing replicate entertainment products is a better choice.

A- 1 Strongly agree (0)
B- 2 Agree (0)
C- 3 Slightly agree (0)
D- 4 Slightly disagree (0)
E- 5 Disagree (0)
F- 6 Strongly disagree (0)

21. Purchasing replicate entertainment products generally benefits the consumer.

A- 1 Strongly agree (0)
B- 2 Agree (0)
C- 3 Slightly agree (0)
D- 4 Slightly disagree (0)
E- 5 Disagree (0)
F- 6 Strongly disagree (0)

22. For me to buy/use counterfeits is convenient.

A- Strongly agree (0)
B- Agree (0)
C- Slightly agree (0)
D- Slightly disagree (0)
E- Disagree (0)
F- Strongly disagree (0)

23. I am motivated to purchase replicate entertainment products because of a special needs such as physiological needs, biological needs, social needs etc.

A- Strongly agree (0)
B- Agree (0)
C- Slightly agree (0)
D- Slightly disagree (0)
E- Disagree (0)
F- Strongly disagree (0)

24. For me to buy/use counterfeits is wise.

A- Strongly agree (0)
B- Agree (0)
C- Slightly agree (0)
D- Slightly disagree (0)
E- Disagree (0)
F- Strongly disagree (0)

Marketing mix

Product, price, place and promotion

25. Generally speaking, replicate entertainment products are reliable.

A- Strongly agree (0)
B- Agree (0)
C- Slightly agree (0)
D- Slightly disagree (0)
E- Disagree (0)
F- Strongly disagree (0)

26. Generally speaking, replicate entertainment products have satisfying quality.

A- Strongly agree (0)
B- Agree (0)
C- Slightly agree (0)
D- Slightly disagree (0)
E- Disagree (0)
F- Strongly disagree (0)

27. On the whole, I am satisfied with my experience with purchasing replicate entertainment products.

A- Strongly agree (0)
B- Agree (0)
C- Slightly agree (0)
D- Slightly disagree (0)
E- Disagree (0)
F- Strongly disagree (0)

28. Overall, my positive experience outweighs **my negative experience** with replicate entertainment products.

    A- Strongly agree (0)
    B- Agree (0)
    C- Slightly agree (0)
    D- Slightly disagree (0)
    E- Disagree (0)
    F- Strongly disagree (0)

29. In general, replicate entertainment products are more affordable.

    A- Strongly agree (0)
    B- Agree (0)
    C- Slightly agree (0)
    D- Slightly disagree (0)
    E- Disagree (0)
    F- Strongly disagree (0)

30. In general, replicate entertainment products are more accessible.

    A- Strongly agree (0)
    B- Agree (0)
    C- Slightly agree (0)
    D- Slightly disagree (0)
    E- Disagree (0)
    F- Strongly disagree (0)

Thank you for your participation. Your real name is not required as part of completing this questionnaire nor be disclosed in future interviews. If you have any questions or comments, please feel free to contact Chris at **krisakinfo@ yahoo.com** or Nana at nana.akaeze@gmail.com, **or call 7189300463.**

Why are some consumers of entertainment products complaisant towards the purchases of replicate entertainment products?

APPENDIX B

Seven Likert-Type Questions

	Strongly disagree 1	Disagree 2	Slightly disagree 3	Neutral 4	Slightly agree 5	Agree 6	Strongly agree 7
1'Academic qualifications	EE-	JH-	SH-2		AD-4	CG-23	PG-19
2'Please indicate the frequency with which you have purchased replicate entertainment products in the past					26	13	9
3'Entertainment is an important part of my family life and activities				1	5	11	25
4'I plan to purchase replicate entertainment products within the next month	14	8	7	6	8	1	4
5'When I purchase products I want to impress others	32	13	2				1
6'When I purchase products it is for my own preference	1	1			5	13	28

7'I think social recognition is important to me	6	1	2	5	12	9	13
8'Work status	6	10	2		5	15	10
9'I became aware of replicate products through information from entertainment reference group or opinion leaders	6	8	2		3	11	18
10'My External sources of influence about replicate entertainment products includes Articles, reviews, advertising, or other activities of the entertainment company	7	1	5		2	19	14
11'My entertainment products buying decisions are mainly influenced by my role and status in the society.	5	13	8		14	8	
12'My Interpersonal sources of influence on purchasing replicate entertainment products includes Opinions of friends, colleagues, relatives, or others	7	2			3	23	13

13'I always attempt to have a sense of accomplishment	1	1		1	3	4	38
14'Age Group		5	1	14	20	8	
15'On average, how much money do you spend monthly on replicate entertainment products?					3	6	39
16'I'm concerned about what other people think of me	22	5			12	2	7
17'I'm self-conscious about the way I do things					9	22	17
18'I have a great personality because I am domineering, aggressive, or self-confidence	15	1	5		1	7	19
19'There's nothing wrong with purchasing replicate entertainment products	7	14	5	5	7	10	
20'Generally speaking, purchasing replicate entertainment products is a better choice	12	13	3	5	6	9	
21'Purchasing replicate entertainment products generally benefits the consumer	1	9	6	5	2	10	15

22'For me to buy/use counterfeits is convenient	22	7	2	5	3		9
23'I am motivated to purchase replicate entertainment products because of a special needs such as physiological needs, biological needs, social needs etc.	18	24			1	5	
24'For me to buy/use counterfeits is wise	24	11	1		2	9	1
25'Generally speaking, replicate entertainment products are reliable	13	19	1	5	8	1	1
26'Generally speaking, replicate entertainment products have satisfying quality	14	4	8	5	9	7	1
27'On the whole, I am satisfied with my experience with purchasing replicate entertainment products	15	8	5		8	12	

28'Overall, my positive experience outweighs my negative experience with replicate entertainment products	14	9	2	5	4	5	9
29'In general, replicate entertainment products are more affordable				5	8	15	20
30'In general, replicate entertainment products are more accessible	1	3	1		11	15	17